George Herman Ellwanger

The Garden's Story

Or, Pleasures and Trials of an Amateur Gardener. Second Edition

George Herman Ellwanger

The Garden's Story
Or, Pleasures and Trials of an Amateur Gardener. Second Edition

ISBN/EAN: 9783337069483

Printed in Europe, USA, Canada, Australia, Japan

Cover: Foto ©berggeist007 / pixelio.de

More available books at **www.hansebooks.com**

THE GARDEN'S STORY

OR

PLEASURES AND TRIALS OF AN AMATEUR GARDENER

BY
GEORGE H. ELLWANGER

Fair Quiet, have I found thee here?
        ANDREW MARVELL—THE GARDEN

SECOND EDITION, REVISED

NEW YORK
D. APPLETON AND COMPANY
MDCCCLXXXIX

COPYRIGHT, 1889,
BY D. APPLETON AND COMPANY.

TO

REV. C. WOLLEY DOD,

MASTER OF GARDENING,

WHOSE WORK AMONG HARDY PLANTS

HAS DONE SO MUCH

FOR THE ADVANCEMENT OF FLORICULTURE,

THIS INCOMPLETE RECORD OF

THE GARDEN-YEAR

IS RESPECTFULLY AND GRATEFULLY

INSCRIBED.

You find me in my garden dress. You will excuse it, I know. It is an ancient pursuit, gardening. Primitive, my dear sir; for, if I am not mistaken, Adam was the first of our calling.—PECKSNIFF.

I am of Opinion that one considerable way to improve Gardening and the Culture of Plants would be to give a description of the Plants themselves; then the Soils, Climates, and Countries where the Plants to be cultivated naturally grow; and what Seasons, Rains, and Meteors they have; which, being imitated as much as possible, perhaps some Plants might thrive better than they do now on the fattest Ground.—PHILOSOPHICAL TRANSACTIONS OF THE ROYAL SOCIETY TO THE END OF THE YEAR 1700, ARTICLE LXXXIV.

PREFACE.

THE publication of a book on the Garden calls for no apology—there are not half enough contemporary works on the subject; there never can be too many. The design of the present volume is to direct attention to the importance of hardy flower-gardening as a means of outward adornment and as a source of recreation. Some of the very many hardy plants, shrubs, and climbers which may be advantageously employed are mentioned, and some hints are given with respect to their use and culture.

I am aware the list is far from complete, even for this rigorous climate, where the line is distinctly drawn by the extremes of heat and cold. To enumerate *all* plants worthy a place under cultivation would require the knowledge and experience of a Loudon; and tastes vary largely as regards the worth and beauty of individual flowers.

It has been the aim to present a simple outline of hardy flower-gardening, rather than a formal treatise or text-book of plants—to stimulate a love for amateur gardening that may be carried out by all who are willing to bestow upon it that meed of attention it so bountifully repays. Nearly all the subjects referred to are such as may be successfully grown in the lower lake region, and, for the most part, have come under notice in the writer's garden.

Different soils and different treatment often produce widely dissimilar results; and even the limited list presented may possibly be found to contain some departure from the well-known types. Moreover, it is pleasant sometimes to look at a flower through different eyes. The flower remains the same, though its perfume may become accentuated, and the garden prove the more inviting the oftener its beauties are set forth.

The following chapters have been so arranged as to present the various aspects of the garden from early spring until late autumn. But the garden year is so interwoven with the many delightful phases of external nature that, the more fully to preserve the sequence of the seasons, it has been deemed advisable to touch also upon the bird and insect life with which it is so intimately connected. The bee, the moth, the butterfly, are all inseparable attendants upon the flowers, and have their mission in the economy of the garden. The birds, also, are constant visitors to every nook and corner, and likewise possess an interest and have a voice in the garden's progress from day to day.

Numerous references to the wild flowers in their native haunts, a chapter on the rock-garden, and a chapter on hardy ferns, have been introduced; and, finally, more or less allusion to the flowers and seasons in literature has been made. The year referred to is that of 1888.

G. H. E.

Rochester, N. Y., 1889.

Contents

	PAGE
PREFACE	v
I. THE GARDEN IN ANTICIPATION	3
II. AN OUTLINE OF THE GARDEN	31
III. THE SPRING WILD FLOWERS	59
IV. WHEN DAFFODILS BEGIN TO PEER	81
V. THE ROCK-GARDEN	105
VI. THE SUMMER FLOWERS	135
VII. TWO GARDEN FAVORITES	165
VIII. WARM-WEATHER WISDOM	193
IX. MY INSECT VISITORS	209
X. HARDY SHRUBS AND CLIMBERS	229
XI. IN AND OUT OF THE GARDEN	245
XII. THE HARDY FERNERY	261
XIII. MIDSUMMER FLOWERS AND MIDSUMMER VOICES	275
XIV. FLOWERS AND FRUITS OF AUTUMN	299
XV. THE LAST MONK'S-HOOD SPIRE	325
INDEX	339

The Garden in Anticipation.

And the spring comes slowly up this way.
                              CHRISTABEL.

Or call it winter, which, being full of care,
Makes summer's welcome thrice more wished, more rare.
                              SONNET LVI.

I.

THE GARDEN IN ANTICIPATION.

IT appears a long way removed still—the goal toward which the lengthening days are slowly trending. In place of rampant *Aries*, ever charging upon the delaying spring, Patience-on-a-Monument would seem an equally appropriate symbol of March, were the signs of the zodiac to be remodeled. The seconds drag through a never-emptying minute-glass, until one wearies utterly of the tedium of the "loaded hours," and wonders not at the impassioned cry of the poet:

O God, for one clear day, a snow-drop and sweet air!

Yet, bluster as he will, March has at most four weeks to retard the " open sesame ! " How gratefully the grass will smile at the first warm rains; and what a caressing odor will arise with the first whiff of *Daphne mezereum*, a foretaste of its sweeter sister, the rosy-cheeked *Daphne cneorum*, and all the train of expectant flowers!

Slowly, yet surely, the hour of the year is advancing. Under the ermine of winter, April's treasures await only the robins' rondeau to call them forth. And what pleasure there is in the anticipation! The swarms of tulips already gathering their forces—the dazzling *rex rubrorums*, the bizarres, and the tall marbled bybloems, which look like the old-fashioned silks of our pretty grandmothers. That bank of oxlips, cowslips, and primroses, too—"crimson-maroon sparkler," "Danesford yellow hose-in-hose," "lilac pantaloons," and ever so many more inviting names—which you placed along the south garden-wall, what a mass of bloom will not push through the mottled earth! And that hamper of daffodil-bulbs, sent by a friend in England, what wealth of beaten gold will not unfold from the fragrant petals!

Will *pallidus præcox* outstrip *obvallaris* in the race; and will "golden plover" vie with "golden dragon"; or can any daffodil, born or yet unborn, excel the glorious bicolor of the Lancashire weaver, John Horsfield? Only, as every rose has its thorn, *Horsfieldi* has its serious drawback, at least with me, in decreasing in vigor every year. Perhaps it is the fault of soil;

more probably a matter of climate. But, inasmuch as I have succeeded in wooing the coquettish *Lilium auratum* so that she smiles instead of frowns, I shall continue to persevere with *Horsfieldi*, which is worth any pains to obtain in the perfect full-blown flower.

To think it has taken all these years to render a daffodil "fashionable"! As if a live flower were a ribbon, subject to the caprice of a milliner! Yet, what may we not expect when lovely woman stoops to blond her tresses, and vandal florists figuratively plunge a flower into the dye-pot? Scarcely a case where beauty is truth, truth beauty. Perhaps, some day, magenta may become the mode, and a magenta gown call for its accompanying flower of the same shade—a chance to let a zinnia scream. The camellia, described in the dictionary as "a genus of beautiful plants," fortunately has had its day—banished with the wax flowers in Wardian cases, let us hope, never to return; too bloodless and too cold even for a chancel; a flower absolutely without a soul. In the *index expurgatorius* should be included the calla lily, which still does lugubrious duty at funerals. Talmage's wish that, when he dies, his grave may be strewed with a handful of violets, a water-lily, a sprig of arbutus, a cluster of asters, rather than that he be laid in imperial catafalque of Russian czar, is a sentiment relatives would do well to consider at the obsequies of those they may be called upon to mourn. The final tributes at the grave, above all, should express the floral preferences of the departed—the old custom of the Indians, clothed in a softer, lovelier garb.

In-door flowers at this season atone, in a measure, for those unobtainable out of doors—always providing one can afford to pay a dollar apiece as the price of a new rose, and shut one's olfactories to the taint of tobacco-smoke and the villainous-smelling stuff shot at the red spider that frequently adheres to the glass-grown queen of flowers. Marie Louise violets and lilies of the valley lose none of their sweetness by being grown out of season. The violet! how pure its wave of fragrance! And the potfuls of "*grand soleil d'or*" and "*grand primo*" *tazzettas!*—surely here is spring incense enough to fill a cathedral at Easter-tide. Is there any odor more delectable than the mingled essence of pineapple, orange, and banana, which this form of the poet's flower exhales? To many, the odor of paper-white (*Narcissus papyraceus*) and the Campernelle jonquil (*N. odorus*) is almost overpowering; they should be used sparingly, therefore—a single spathe will suffice to scent your library. Powerful enough they are to have pleased Baudelaire, who, preferring musk to violets or roses, declared, "My soul hovers over perfumes as the soul of others hovers over music." There is, indeed, an intoxication, and often a strong association, in the subtile odor emitted by certain flowers. Does not the perfume of *Lilium auratum*, stealing from the spotted petals, recall the reedy jungle and the stalking tiger? Or a gorgeous epiphytal orchid, steeped in its mysterious perfume, does it not simulate unconsciously some strange form of tropic insect or animal life? I oftener recall a flower by its odor, to which sentiment tenaciously clings, than by mere characteristics of form or color. What an indelible aroma, that of the fragrant everlasting of the fields!—a wild, haunting odor, as of fallen leaves after the latter rains, when the sun extracts their essences, rather than the characteristic fragrance of a flower. Through its rustling, ashen petals I already inhale the autumn from afar, and anticipate the last sad cricket's cry. If Addison be taken for authority, we can not have a single image in the fancy that does not make its first entrance through the sight *—a dogma which,

---

* "On the Pleasures of the Imagination."

though emanating from the "Spectator," is manifestly sophistical and untrue. Was Addison deficient in the sense of smell (the voice of a flower); or was a thrush's song powerless to awaken in him a sentiment of sublimity? But Addison does not mention odors, and, for the most part, I take it, did not like external sounds; or was it Steele who wrote the essay "On the London Cries"?

Bulwer declares, the only perfume a man should use is soap and water—a heresy. I would not for a moment commend musk, or even ylang-ylang; though the latter, it seems to me, is preferable to the compound of Jean Maria Farina with which men fairly saturate themselves. Consider its ingredients: orange, cedrat, neroly, bergamot, and rosemary—scent enough to trap a cougar. But this is supposed to be fashionable; while a hem-stitched handkerchief, with a lingering scent of violets, has no right to peep from the masculine pocket. Why should everything dainty be monopolized by the fair sex? Has it not enough, with its feathers and ribbons and laces and jewelry, without carrying out the adage to its ultimatum, "sweets to the sweet"? It even robs masculinity of any proprietorship to color, except what little can be focused into a scarf, or polka-dot a waistcoat.

To be sure, there are those striped Joseph's-coats one meets at the sea-side, appropriately termed "blazers," which woman openly professes to admire, only to contrast them inwardly with the sea-side habiliments of her own human form divine. Even her blue bathing-dress she has deliberately pirated from the sailor of the high seas, and pilfered the crowning characteristic that proclaims man a man—the stovepipe hat.

Let those of the sterner sex who love the delicate aroma of a flower not hesitate to use its essence when distilled by an Atkinson, if the flower itself can not be had to take its place on the lapel. Does not Dumas *père*, in the "Vicomte de Bragelonne," speak of the Bishop of Vannes as exhaling "that delicate perfume which, with *elegant men* and women of the grand world, never changes, seeming to be incorporated in the person of which it has become the natural emanation"? Another case where they manage these things better in France. It is well known, moreover, that flower-essences are prophylactic and antiseptic—the more reason why they should be employed, in moderation, and that their use be not monopolized by woman. "There are perfumes," says Gautier, "which are fresh as the skin of a child, green as spring meadows, recalling the flush of sunrise, and carrying with them thoughts of innocence. Others, like musk, amber, benzoin, spikenard, and incense, are superb, triumphant, mundane, provocative of coquetry, love, luxury, festivity, and splendor. Were they transposed to the sphere of colors, they would represent gold and purple."

I open the jar of rose *pot-pourri* to flood the room with the subtile essence of June. No evanescent odor, but one that permeates and clings, evaporating not, changing not its sweetness from year to year. I do not refer to the dry, soapy-smelling article of commerce labeled "Tea-rose *Pot-pourri* from Japan," but to the old-fashioned "rose-jar," made from your own garden-roses, blended with a sufficiency of other sweets to hold its perfume immutable. It is difficult to give a precise recipe for a rose *pot-pourri*, for no two ever turn out quite alike. I would say, however, with fat old Baron Brisse in the preface to an *entrée* in his " Petite Cuisine ": "There is a certain point in this preparation rather difficult to seize; but this is the way to set about it in order to be complimented:"

The roses employed should be just blown, of the sweetest-smelling kinds, gathered in as dry a state as possible. After each gathering, spread out the petals on a sheet of paper and leave until free from all moisture; then place a layer of petals in the jar, sprinkling with coarse salt; then another layer and salt, alternating until the jar is full. Leave for a few days, or until a broth is formed; then incorporate thoroughly, and add more petals and salt, mixing daily for a week, when fragrant gums and spices should be added, such as benzoin, storax, cassia-buds, cinnamon, cloves, cardamom, and vanilla-bean. Mix again, and leave for a few days, when add essential oil of jasmine, violet, tuberose, and attar of roses, together with a hint of ambergris or musk, in mixture with the flower ottos to fix the odor. Spices, such as cloves, should be sparingly used. A rose *pot-pourri* thus combined, without parsimony in supplying the flower ottos, will be found in the fullest sense a joy forever.

Notwithstanding the rarity of flowers at this season, no one with space enough for the smallest kitchen-garden need be without at least an abundance of violets. A small stock of strong young plants, placed in good soil in May in a partially shaded position, will have increased sufficiently by November to supply a hot-bed. These should be planted within a few inches of the glass, early enough to insure their rooting well before extreme cold weather. The hot-bed should be placed in the most sheltered and sunshiny position, and be thoroughly protected on the sides with leaves or straw, and the sashes covered with thick matting and boards to exclude frost. So soon as the weather allows, in spring or during the winter, air should be given gradually during the day, recollecting that cold currents of air should be guarded against. As the weather becomes warm, and the plants require it, they may be watered occasionally. Pinching back the runners will increase the bloom. After blossoming, lift the plants, divide them and place them in the open, as before. During extreme dry weather they will naturally be much benefited by an occasional watering and mulching. No one who cares for flowers will grudge the little trouble and trifling cost of a violet-bed which yields its wealth of blossom when other out-of-door flowers are still buried beneath the snow. I know of nothing that affords so much satisfaction for so little pains. Marie Louise is incomparably the most fragrant, floriferous, and satisfactory variety for hot-bed culture.

From the adjoining hill-side at nightfall I hear the weird nocturne of the small screech-owl. A pair has always had its abode in the covert, in company with the red squirrels that bark so fiercely at the falling nuts in autumn. They each give an air of wildness to the surroundings, and one feels as if the trees had found an expressive voice. I can not comprehend why the owl should invariably be associated with gloom and deeds of evil, or that his voice should allow us to forget for a moment his accomplishments as a mouser. When other birds have deserted us, and even the squirrel remains in his hollow tree, the cry of the owl rings out sonorously on the winter twilight, "I am here!" Well may Thoreau rejoice that there are owls, and Jesse admire their soft and silent flight. Charles Lotin Hildreth is superlatively the poet-laureate of the bird of wisdom. Shakespeare, Barry Cornwall, Shelley, Wordsworth, Jean Ingelow, and Tennyson must each and all give place to his apostrophe. Take the opening and the closing stanzas, for instance :

> There is no flame of sunset on the hill,
>   There is no flush of twilight in the plain ;
> The day is dead, the wind is weird and shrill ;
>   Amid the gloom the sheeted shapes of rain
> Glide to and fro with stealthy feet and still,
>   And, wilder than the wood's autumnal moan,
>   A voice wails through the night, "Alone, alone!"
>
>   . . . . . .
>
> Night deepens on the haggard close of day
>   With wilder clamor of the wind and rain ;

> Louder the beaten branches groan and sway;
> And fitfully the voice comes once again,
> Across the fields, more faint and far away.
> Is it the dark bird's wailing backward blown,
> Or my own heart that cries, "Alone, alone!"

The snow is fast retreating despite the raw March winds, though St. Patrick and the vernal equinox have yet to engage in their accustomed brawl. Indeed, St. Patrick never comes in without brandishing his blackthorn. As 'tis an ill wind turns none to good, so the dreaded equinoctial is not without its advantages. Not having Blasius for authority, I can not tell why it is so; nevertheless, the weather-vane of the equinox for the three days of its duration is an index to the character of the weather for the succeeding two or three months. A puzzling rule of three, no doubt, but why not as probable as that three consecutive white frosts are a never-failing sign of rain? To be more explicit, the general direction of the wind and character of the weather during the several equinoxes would seem to be followed during the greater portion of the next quarter of the year by a like general direction of the wind and character of weather.

Avant-couriers of spring continue to blossom diurnally through the post, in the shape of flower and vegetable catalogues. These unfold some interesting studies in form, and reveal new possibilities of color. Many of the covers seem Koula rugs transformed into card-board; and the hideous greens, saffrons, and magentas that gape from the Anatolias in the carpet-store windows appear to have been lavishly borrowed to heighten the effect of the foliage and fruit of some new strain of gourd, ruta-baga, or colossal onion. The most powerful appeal of the season is a full-page plate of liver-colored tomatoes and zinnias in combination. In another distinctly æsthetic overture, a plant of the *Ipomæa* tribe, sent out under the name of moon-flower, has embowered an entire cottage; while the moon itself, represented as rising in the horizon, shines only with a borrowed splendor in the presence of this high-class luminary. When the catalogue informs one, in addition, that "the flowers, when unfolding, expand so rapidly as to be plainly seen, affording amusement and instruction, and that, being a free bloomer, the effect on a moonlight night is charming," the reader need no longer doubt the advent of the floral millennium.

Surely it is the weather that the crows have been denouncing so vehemently for several evenings from their roost in the immediate vicinity. If we have not the rook, I am glad we have his larger Plutonian cousin. His dusky shadow and husky bass have a charm of their own, and harmonize with the bleakness of early spring and the somberness of late autumn. Apart from the pestiferous English sparrow, the crow supplies almost our only winter voice. I place him with the black hellebore or Christmas-rose—a very good thing to have until there is something better to take his place. The Ettrick Shepherd should have substituted the crow-blackbird when he said, "The crow is down in the devil's book in round hand." I am glad to hear Phil Robinson say he should be reluctant to deny this bird every one of the virtues; and John Burroughs exclaim: "I love him; he is a character I would not willingly miss from the landscape."

The advance-guard of the robins has come, behind its usual time, but their reception has been too cold as yet to expect them to proclaim their presence in an audible manner. For the robins' silence the sparrows are doing double duty. I shall have to set my long pole in motion again, and banish them from the front verandas to those of my neighbors. Birds, it is well known, will not endure being disturbed from their roosts; and one or two dislodgments after nightfall will suffice to rout even the sparrow, although he is so disgustingly numerous that there are soon others ready to take his place as public defilers. Too cunning to be poisoned, a light charge of No. 12 shot is the best means of allaying his obstreperous cry. I usually leave the corpses of the chief offenders, the noisiest among the cock birds, in some conspicuous place for a day or two, and the matutinal tom-tom in the sugar-maple near my sleeping-room gives place to a sense of delicious repose. One of the necessities of the hour is a noiseless powder, and a practical sparrow-gun, light and cheap enough to be generally utilized. A twelve-gauge gun answers the purpose, save for the loudness of the report; and a small rifle is effective, but the successful use of this requires too much skill to meet the popular demand. Through the means proposed, no one need be disturbed except the chief offender, and a liberal supply of cartridges would perceptibly rid one of his loathsome presence. "The sparrow carries no purse," says Phil Robinson, "for he steals all he wants; and his name is in no directory, for his address is the world." If Bryant lived to-day, he would assuredly change his false refrain, "The Old-World sparrow is welcome here." An anonymous writer voices a charming sentiment : "Cursed be the man—the enemy of the peace of all civilized Americans—

who imported them! He should be incinerated, and his ashes blown by the four winds to the four quarters of the globe."

The melancholy fact exists, notwithstanding, that the sparrow is here, and stands urgently in need of a prescription. He has succeeded in his dual *rôle* of harassing both mankind and his own desirable congeners. One by one he has driven away the song-birds from our homesteads, appropriating the nesting-places of the wren, the bluebird, and others, whose presence was invaluable in the orchard and among small fruits. The oriole still remains, concealed by the apple-bloom, or comparatively secure from assault in his rocking-chair in the elm. The song-sparrow and purple martin are diminishing yearly, the robin, blackbird, and oriole alone being able to resist his malicious persecution. In like manner, the Rocky Mountain trout has been placed in streams tenanted by the speckled trout, only to override and diminish a superior game-fish. Can not the champions of the English sparrow people the country with the Australian rabbit, or disseminate the Swiss goitre, as an act of philanthropy? A State or Government bounty on the sparrow's head would help to thin his polygamous brood; his slaughter for "potted game" would add largely to the score;

the sportsman's trap might ingulf him by the thousands; and wholesale netting, as practiced abroad, would well reward those who supply the restaurant larder. The shrike, or butcher-bird, is an admirable *matador* so far as he goes, and would, no doubt, end in exterminating him, with man's assistance, did he exist in sufficient numbers. Let us pray, meanwhile, for the advent of a sparrow-bug, or *Passer-aphis*—some insect-scourge such as besets the inanimate world—to aid in delivering us from this feathered Philistine.

The chimney-swallows, which last summer awakened me with their chattering and whirring in the chimney, at all times of the night and early morning, will trouble me no more. A wire screen placed across the top of the chimney has rendered a little folding of the hands to sleep possible at five in the morning. The chorus of the *Hylodes*, or peepers, is yet in store—that piercing treble launched against the quiet nights of early spring that nothing—even the katydid—can equal in strident intensity, and that no earthly power can still. Fancy attempting to go to sleep in a country house near a swampful of these shrieking demons! "It is a plaintive sound, a pure spring melody," says Burroughs, for once apparently forgetting himself, or led astray by its association with spring. But he sets us comparatively at ease by stating that there is a Southern species heard, when you have reached the Potomac, "whose note is far more harsh and crackling. To stand on the verge of a swamp vocal with these, pains and stuns the ear. The call of the Northern species is far more tender and musical." It is at least some consolation to know there are others worse off than ourselves.

The uproarious crow-blackbird, too, is sure to return to the Lombardy poplars in April. A volley of coarse shot alone can drown his discordant gutturals, which he hurls at you in utter disregard of the exquisite sensitiveness of the human tympanum. Fortunately for mankind, he is less numerous than the nauseous sparrow, or deafness were necessarily the all-prevailing malady. How many of these oral miseries there are awaiting us! It is enough to develop a gouty diathesis to think of them.

The blue-jay is almost always referred to as the most discordant among the birds, while little fault is found with the harsh voice of the grackle or crow-blackbird. Compared with the latter, the jay is a paragon in manners, dress, and every characteristic, unless we except his habit of pilfering the nests of his neighbors. His voice certainly has a meaning from his vantage in the tree-tops. It is emphatically a cry of warning, uttered loud enough for every feather in the forest to hear, that an enemy is intruding upon the sacred domain. His crest of sapphire would atone for his shrill clarion, were not the meaning of his cry a sufficient excuse in itself. The grackle, on the other hand, only screams incessantly to hear himself scream, and to drown the voices of the song-birds.

In Harris's "Treatise on Insects injurious to Vegetation," the crow-blackbird is made to pose as a public benefactor. The reader, at first shocked by the statement that "few persons, while indulging in the luxury of early green peas, are aware how many insects they unconsciously swallow," is somewhat relieved later on by being told that these "buggy peas" contain, in the first instance, a minute whitish grub, which larva is changed to a pupa within its hole in the pea in the autumn, and before spring casts its skin again, becoming a beetle (*Bruchus pisi*), only to fly out into the awaiting maw of the crow-blackbird! "Buggy peas," I admit, do not sound appetizing at first hearing; still, were we to draw the line at such trifles, I fear our vegetable diet would necessarily be greatly restricted. So long as we eat the insects with the special vegetables they infest, there can be but little objection. The strawberry and raspberry parasites are, certainly, exceptions; for no one could taste and swallow more than one of either, and live to tell the tale. The mushroom-worm, the cabbage-louse, the lettuce-hopper, the Brussels-sprout thrip, and dozens of other jumping, wriggling things which the cook sends to table, possess invariably the exact flavor of the several vegetables they garnish. It would only be by serving the wrong or foreign insects with a particular dish that any gastronomical syncretism could result. The *argumentum ad gulam* advanced for the existence of the crow-blackbird is, therefore, untenable, and I fail to discover any excuse for allowing him to usurp the place of the starling, with whom he is forever quarreling.

Another blanket of snow has been heaped upon us, just as the previous vestiges had disappeared and there were hopes of an end to the interminable hibernation. It was a halting philosopher who termed snow the poor man's manure, for want of a proper definition. The ammonia it contains one might better be without at this season, when every shrub, plant, and grass-blade is crying for the caress of the rain. Apparently the snow came from the asperous east. I have wondered why the east wind should be so unkind, coming, as it does, from lands sentient with sunshine and steeped in tropic warmth. A wind like Ruskin's "plague-wind, made of dead men's souls—such of them as are not gone yet where they have to go, and may be flitting hither and thither, doubting themselves of the fittest place for them." I find the east wind has been grossly maligned; it is the west wind that bears the venom of Boreas and the stratus-cloud in its icy breath, surging on an upper current of the atmosphere, and coming only in appearance from the east on a counter under-current of air. The Rocky Mountains are the real seat of the dreaded "easterly" storm, and they—not the east wind—deserve our strictures.

In point of viciousness and duration the present equinox exceeds any other I have known. The chanticleer on my neighbor's house-top has been whisking seemingly from each point of the compass at once; and every variety of weather, from an east wind bitter as quassia to the most brutal of westerly blizzards, has raged unremittingly for six days. I defy even Sir Admiral Fitzroy to forecast the weather from so heterogeneous a horoscope—a combination of winds that has blown evil to me and good to my allopathic doctor, whom I shall exchange for a homœopathist if I survive to undergo another vernal equinox in this latitude. Without a word of warning, I awoke in the night with the sensation of having been pounded in a mortar, and with a Spanish chestnut-burr sticking in my throat. I never realized before what the innocent-looking yellowish mixture was that he prescribed for the children—potash and iron—with which he has been deluging me almost hourly, night and day. A doctor ought to be exiled for forcing such revolting stuff upon helpless patients—a remedy which is almost worse than the disease. Hugh Miller's "Testimony of the Rocks," or Borden Bowne's "Studies in Theism," would be a delicious lenitive, in comparison. If I live, I shall find out whether his statement is true: that it is the great catholicon for diphtheritic and laryngeal troubles, and that nothing else can disperse the dusky spots on one's throat, or cause the white ones to "exfoliate," as he pathologically expresses it. It was an exhilaration to me, with all sense of taste and smell temporarily destroyed through his vile prescription, to learn that he had an allopathic doctor under treatment, and was dosing him in the same wholesale manner that he was medicating and mending me.

I shudder when I think of the books I ought to "assimilate," directly and indirectly, in connection with the subject of gardening. Think of it! Darwin's "Vegetable Mold and Earth-Worms," Dyer's "Natural History of a Flowering Plant," Harris's "Talks on Manures," Warder's "Hedges and Evergreens," Darwin's "Climbing Plants," Berthold Seemann's "Revision of the Natural Order Hederaceæ," Bentham and Hooker's "Genera Plantarum ad Exemplaria imprimis in Herbariis Kewensibus servata definita," Loudon, Downing, Lindley—there is, apparently, no end to them.

The illustrations of the unabridged dictionaries, too, that one is forced to encounter; the cuts of the snakes and the reptiles that are coiled around every other page of a book one is compelled to read! One always opens the dictionary at the snake pages, or is confronted with a growling peccary, a hooded basilisk, a *Mephitis Americana*, or open-mouthed crocodile, to promote a shiver that is liable to develop into bronchitis. The most barbarous words, likewise, seem always placed at the top of the page in staring capitals—medical and scientific terms one must perforce swallow, even though the dose be nauseating. The serpent and lizard appear to be the favorites of the lexicographers; I find no cut of the fruit which tempted Eve. As to the flower and tree illustrations, the representations that have served to portray *Lilium bulbiferum*, the magnolia, and the weeping-willow, are past praying for. They have all done valiant duty, and deserve to be placed upon the retired list with a liberal pension.

The gardener has just called, bringing the cut flowers and his summary resignation at the same time. He, too, has caught the spring fever and desires a change. How we will miss his Brussels-sprouts and endive. He was worth having for his success in vegetables alone; he knew enough to cut asparagus close to the ground without being told, and his heads of cauliflower were so delicious *au gratin!* What is to become of all the spring work, meanwhile, which comes upon one so suddenly when it finally does come? Think of all the flower-borders that have to be uncovered, the leaves to be raked up and carried to the pile of leaf-mold, the spring-manuring and spading, the seed-sowing and pruning, the lawn-raking and rolling. and the general cleaning and overhauling! If I take a hand in it myself, there is always the danger of catching cold, and not for worlds would I undergo another medication. I must get Casper, the former German gardener, back again.

What if he did let the greenhouse plants become a prey to the red spider?—he was always so good-humored, and accomplished so much! Your short, burly, broad-backed gardeners somehow always work easily and quickly; they have not so far to bend over; the weeds jump up to them, and they handle a rake as if it loved them. A Mecklenburger for work and an Irishman for blarney.

The long-awaited change has come at last—the promise the wild geese flying north chorused from the upper air. Song-sparrow, bluebird, meadow-lark, plover, and redwing have dropped down suddenly and simultaneously from the sky, and from the swamps I hear the croaking of the frogs, eager to drink of the first warm rain. The scarlet maple (*Acer rubrum*) is bursting into bloom, and one can almost see the grass sprouting, so thirstily does it absorb the moisture. The woodcock have already returned to their summer haunts; I found them on the dry knolls March 25th. Referring to the record of the past eighteen years, the tables show that an early spring occurs about every other year in this vicinity. In 1880 the robin appeared February 27th; the bluebird and song-sparrow a day later. In 1874 and 1878 they appeared simultaneously March 3d, compared with March 27th, 28th, and

29th in 1885. In 1880 and 1882 the scarlet maple blossomed March 2d; in 1872, 1873, and 1879, April 10th; in 1885, April 20th. The earliest pipings were sounded from the marshes March 9th in 1877 and 1878; the latest, April 20th in 1885 and April 14th in 1874. The earliest high temperatures recorded were those of March 31, 1875, 69°; March 27, 1882, 64°; April 23, 1885, 90°.

Compared with previous seasons, therefore, the present has been no worse than the average. One must needs be grateful for the meager pittance March doles out in the way of blue skies and stray shadows on the garden dial. The last few days of the month have been prodigal of sunlight; and see, the urn of the first crocus already flaunts its hoarded gold to herald the returning flowers of spring!

An Outline of the Garden.

Every wyse man that wysely would learn anything, shall chiefly go about that whereunto he knoweth well that he shall never come. In every crafte there is a perfect excellency which may be better known in a man's mind than followed in a man's deede. This perfectnesse, because it is generally layed as a broad, wyde example afore all men, no one particular man is able to compasse.—ROGER ASCHAM.

II.

AN OUTLINE OF THE GARDEN.

ROGER ASCHAM might have alluded to gardening when he penned the foregoing lines. The art of gardening is comparatively easy in theory; its consummation is more difficult in the soil. And it is not unlikely we shall find the author of "The Scholemaster" easier to read between the lines than appears at first sight, in the interval that shall elapse between the matins of the first snow-drop bells and the vespers from the last monk's-hood spire.

I write of the hardy flower-garden. This may be large or small, though, beyond a certain indispensable area, its perfected beauty depends not so much upon mere size as upon intelligent treatment. A small plot properly laid out, judiciously planted, and kept in finished order, will produce more satisfactory results than ten times the space poorly cultivated and insufficiently maintained, It is essentially a garden maxim, that whatever is worth doing at all is worth doing well. So that, first of all, the grounds should be no larger than can be properly looked after. Grass-grown walks, untrimmed edgings, a lawn run to weeds, at once proclaim the untidy gardener, and detract from the best efforts of the flowers themselves. I do not speak of the stiff, formal garden, divided into methodical squares, where everything must be equally balanced; or of "bedding-out," "carpet," or "ribbon" gardening. I speak of the hardy flower-garden, where, in its effect as a whole, a sense of tidiness combined with natural grace of outline and harmony of grouping should prevail. If the space be too large to be perfectly maintained, diminish it; but let whatever space there may be under cultivation suffer no neglect or show no parsimony of care.

No arbitrary rules will suffice to produce a perfect garden, for, in the very nature of things, no two gardens can be just alike. Each one should seek his own expression in the combination he strives for. For this there exists infinite variety of material, adaptable to the particular soil, exposure, and character of the space one would adorn and idealize. A charming feature

of one garden may not be allowable in another,
either through lack of space, difference of exposition, or natural incongruity. Thus, a miniature pond for the cultivation of bog-plants—a
delightful feature of the garden where it may be
carried out—can not be introduced with propriety on high exposures. Nor can a bank of ferns
be placed to advantage where they have not the
coolness and shade with which they are associated, and without which they can not be satisfactorily grown. In a large place, possessing the
resources of abundant shade and variety of surface, there are few desirable effects which can
not be produced. Here the landscape-gardener
proper has a field for the practice of his art, and
the proprietor an occasion for the gratification of
his taste. In small grounds, however, as distinguished from the large estate, one need not be a
Crœsus to enjoy the pleasures of gardening.
There is force in Bacon's statement: "A man
shall ever see that, when ages grow to civility
and elegance, man comes to build stately sooner
than to garden finely; as if gardening were the
greater perfection." And yet, with the wonderful advancement of the arts in this country during the latter part of this century of progress,
the art of gardening, it must be admitted, has
also shown marked improvement. Occasionally

we find those who are content with a geranium-bed as the means of outward embellishment; more often a finer perception of external adornment is manifested, though Bacon's statement remains apposite to-day. The objectionable forms of gardening, however, are being superseded by a more natural style—a revival of the old-fashioned hardy flower-borders, masses of stately perennials, the hardy fernery, the rock and bog garden, the azalea and rhododendron beds. Poor indeed is the city veranda which has not its *Clematis Jackmani* to flutter the purple of royalty, and lonely the door-yard without its clump of madonna lilies or perpetual roses. A comparison of the flower-catalogues of to-day with those of fifteen years ago shows beyond contradiction the advancement of the cultivation of hardy plants. Notably the case with new varieties of roses and flowering shrubs, progress is also observable with perennial flowers. The tendency of the age to cast aside poor for better forms, to resurrect or improve the old, includes the flower-garden among the many things to feel its quickening influence. Material we have in abundance; it only remains for us to utilize it and adapt it to the ends in view. To create the ideal in landscape floriculture, to surround ourselves with lovely forms of nature, with no discords to jar upon the sense of harmony, can only be attained by carrying out the suggestions of nature itself, applying them with all their possibilities of modification, change, and adaptation to the means we would attain.

From the first, all appearances of stiffness and rigidity of outline, whether of walks, roadways, or borders, will be studiously avoided. The natural line of beauty we should attempt to reproduce. The placing of ornamental trees and shrubs will depend on the situation and exposure; the arrangement and grouping of flowers and foliage-plants, on one's sense of color and correct interpretation of effects.

I like the hardy shrub border, the low-growing and comparatively less robust shrubs, for a screen next the highway; for no garden, I think, can be satisfactory without privacy. Glimpses of the interior may be afforded the passer-by, but retirement and shade constitute two of the greatest charms of the garden. The hardy shrub border combines privacy and beauty. In it I would have, among others, for the larger subjects, the Japan quinces; many of the *Deutzias;* the common barberry, for its colored fruit in autumn; the purple-leaved, for its effective foliage; the light-colored althæas, or rose of Sharon; the *Calycanthus*, or sweet-scented shrub, for its fragrance; the large-flowered and changeable hydrangea; the dwarf and golden-leaved syringas, or mock-orange; the double-flowering *Prunus;* the spiræas in variety; the fragrant *Ribes,* or yellow flowering currant; some of the smaller lilacs; the dwarf sweet-scented *Magnolia Halleana;* the *Exochorda,* the *Daphne mezereum,* the variegated dogwood, the white *Weigela,* the purple-leaved plum, the cut-leaved sumac, the golden, fern-leaved, and cut-leaved elder.

Such strong-growing subjects as the *Forsythias,* the large magnolias, the snow-ball, and the lilac are apt to domineer over their companions. If possible, they should be placed by themselves where they may have abundance of room to develop their full beauty. Even with most of the less robust examples I have enumerated, the pruning-knife must be applied at the proper season, to keep them shapely and within bounds. It should always be remembered that each shrub has its characteristic habit of growth. This should be retained, so far as possible. To trim all shrubs alike, is to ruin their beauty and mar their natural grace of outline. Judicious pruning, to promote health and vigor, is necessary. Old growth requires thinning out occasionally, and obtrusive root-sprouts and suckers need to be removed. Althæas, spiræas, lilacs, and honeysuckles may be trimmed early in spring. *Deutzias, Forsythias,* mock-oranges, and *Weigelas* flower on the wood of the preceding year's growth, and should be pruned after June flowering, when the old wood may be shortened or cut out. Evergreens may be pruned in April or May, to thicken the growth and preserve shape. Happily, the practice of shaving trees and shrubs, the art of "verdant sculpture," is less common than formerly. Legitimately used to assist Nature, the pruning-knife becomes a valuable assistant; too often it is the means of destroying identity of form.

Of the scores of *Weigelas* or *Diervillas* under cultivation, I know of few to be recommended for the choice collection of hardy shrubs. For the most part the flowers are of a displeasing color, while the shrub takes up a large space which, with the "rose-colored" kinds, might be occupied to far better advantage.

For the dwarfer shrubs and plants of the hardy shrub border I should employ such subjects as the tree and herbaceous pæonias, the large perennial phloxes, the two forms of the Japanese anemone, and some of the taller lilies. The tall and hardy reed-like grasses—*Erianthus ravennæ, Eulalia japonica* and its varieties—may be used with good effect in both the shrub and flower border, though still more striking by themselves.

The width of the shrubbery border should depend upon its length, and also upon the extent of space between the highway and the residence. Very wide borders, where the frontage of a place is not deep, contract the grounds and curtail the expanse of lawn. Judicious planting becomes the more necessary the wider the border, or large patches of bare ground will obtrude. Very long, narrow borders are equally objectionable, and have a stinted look.

I take it for granted the lawn will receive the consideration it deserves. Undoubtedly the most important element of beauty of the grounds, without it all other forms of embellishment go for little. Green is the natural relief of floral color; and in no way does floral color stand revealed so vividly as when set off by a perfect sward. To form a perfect lawn,

> ful thikke of gras, ful softe and swete,

requires pains. The soil must neither be too poor nor too rich, but contain a sufficient depth of good garden soil to insure against drying out during hot weather. Above all, earth removed in excavating, usually composed of clay or gravel, should never be used for surface soil, as is not unfrequently the case. Jealous guard should be kept, when sewer or other excavations are made at any time, that the subsoil be not left upon the surface, or dry grass patches will invariably show themselves with the first hot weather.

With what grasses should the turf be formed? This has been answered a great many times in a great many ways. Assuming that the sower knows precisely what kind of seed he would sow, the difficulty arises of procuring pure seeds of the species desired; the only sure way is to have the seeds tested by an expert. I quote two authorities on the best grasses for the lawn.

W. J. Beal: "Two sorts of fine *Agrostis* are sold under the trade name of Rhode Island bent, and, as trade goes, we may consider ourselves lucky if we get even the coarser one. The finest, a little the finest—*Agrostis canina*—is a rather rare, valuable, and elegant little grass, which should be much better known by grass farmers as well as gardeners than it is. The grass usually sold as Rhode Island bent is *Agrostis vulgaris*, the smaller red-top of the East and of Europe. This makes an excellent lawn. *Agrostis canina* has a short, slender, projecting awn from one of the glumes; *Agrostis vulgaris* lacks this projecting awn. In neither case have we in mind what Michigan and New York people call red-top. This is a tall, coarse native grass, often quite abundant on low lands, botanically *Agrostis alba*. Sow small red-top, or Rhode Island bent, and June grass (Kentucky blue-grass, if you prefer that name), *Poa pratensis*. If in the chaff, sow in any proportion you fancy, and in any quantity up to four bushels per acre. If evenly sown, less will answer; but the thicker it is sown the sooner the ground will be covered with fine green grass. We can add nothing else that will improve this mixture, and either alone is about as good as both. Under no circumstances sow a little oats or rye 'to protect the young grass.' Instead of protecting, they will rob the slender grasses of what they most need."

Daniel Batchelor: "As to the grasses best adapted to soils and situations, it may first be said that a wet soil is hardly to be considered as a fit situation for a lawn; nevertheless, there are places where a moist condition of the soil can not well be avoided, and for such the best grasses are *Poa trivialis*, or rough-stalk meadow-grass; *Alopecarus pratensis*, or meadow fox-tail; and *Agrostis vulgaris*, or red-top. For average good soil I have had the best results from a seeding, in about equal proportions, of *Poa pratensis*, or Kentucky blue-grass; *Festuca duriscula*, or hard fescue; *Agrostis canina*, or creeping bent;

Cynosuris cristatus, or crested dog's-tail; and *Lolium Paceyi*, or Pacey dwarf rye-grass. The two last named are especially adapted to light, dry soils, as they are deep-rooted and very fibrous, and will continue green in the driest of weather, even when the Kentucky blue is apparently dead. I may here state that there are hundreds of bushels of English rye grass-seed imported and sold for Pacey's dwarf rye, but it is a cheat, as the former is not hardy in our climate. Pacey's is quite a hardy variety, and is, I think, of Scottish origin; at any rate, it is one of the best grasses for either lawn or sheep-pasture.

"Some persons recommend, in mixtures, such grasses as the *Festuca ovina*, or sheep's fescue, and the *Festuca tenuifolia*, or slender fescue. I think that both of these are objectionable on fine lawns, as they grow erect and tufty, while the leaves are round, wiry, and sedge-like; the color, too, of both is blue, especially that of the slender fescue; and, altogether, these grasses look intrusive and patchy when contrasted with the flat, ribbon-shaped foliage of those I have ventured to name with approval."

The addition of white clover to whatever grasses one may sow is a matter of individual preference. On some light soils it is a most valuable adjunct, if not a necessity, and many would not be without its sprightly presence. But of whatever grasses the lawn may consist, the necessity of drainage in low situations, and thorough preparation of the ground in all cases, will be readily conceded. It is only in good, well-drained soil that the finer grasses will remain verdurous during the intense heats of midsummer. Spring is doubtless preferable to autumn sowing, still, in cases where it can be done, it is a positive advantage to prepare the ground in autumn, and allow it to settle thoroughly through the winter. The addition of a small per cent of lime at the outset is to be recommended, except on thin, sandy soils. These should be fortified with a liberal supply of old manure and good loam and muck, with the addition of a sprinkling of quicklime. Thorough rolling previous and subsequent to seeding is of prime importance.

Once formed, it is a common error to suppose the lawn will take care of itself. A top-dressing of fine compost or some good commercial fertilizer should be applied at least once every other year early in the spring. Either is preferable to manure of any form, which is unsightly. Fresh manure is especially to be avoided, if for no other reason than the crop of weed-seeds it contains. Inequalities of surface should be filled up with loam and freshly seeded, and the roller be thoroughly applied over the entire surface after raking. Always sow grass-seed liberally. It is a mistake to leave either a close or a heavy sward over winter. Cut too short, the grasses suffer; left too long, they are unsightly and start slowly. The lawn should not be shorn closely or frequently the first year; after that, frequent mowings are advantageous where the shorn grass is left to enrich the sward. Often the sweep of the lawn is spoiled by being too closely planted with trees or shrubs, frequently with both; or by being cut up with flower-beds. While some shade is very desirable, too much shade is injurious to the growth of grasses; and close planting interferes with the sense of generous breadth which the lawn should impart.

Even with all possible pains and precautions we are still without a perfect lawn. The grand army of weeds remains to be combated—perennial pests innumerable; annuals which sow themselves a thousand-fold; plantains voided by granivorous birds; purslane traveling on wings of the wind; dandelions, rooted deeply as ingratitude; sorrel, lover of sandy soil; mouse-ear chickweed; yarrow, daisies, mosses, lichen;

and that English sparrow among the weeds, crab-grass, whose maw is insatiable and whose worm never dies—all these fail not to appear at their appointed time. Persistent warfare with the gouge-knife is the only means of keeping down the perennials. The spreading, self-sowing annuals that creep along stealthily, undermining the grasses, are less amenable to treatment, and frequently require to be dug up in patches, resodding or resowing the spots whence they were removed. All of these pests, unpleasant as they are, we would willingly exchange for the crab-grass (*Paspalum sanguinale*), the bane of American lawns. This annual appears most disagreeably during August, at the time of its inflorescence, its brownish stems rising from large tufts, which crowd out the finer grasses, giving the sward the appearance of having been burned, and utterly ruining the appearance of the turf wherever it obtains a foothold. It revels in drought and hard-pan, and, like the horse-leech's daughters, cries out continually, "Give, give!" Practically inexterminable, the mower passes over its wiry stems, which cling to the ground and perfect their seed-sowing for another year. Good soil, abundant watering, and shade alone tend to check it. The only thing to be said in its favor, as distinguished from other lawn pests, is its late appearance and comparatively short duration.

Frequently ants and the white grub—the larval grub of the May-beetle (*Lachnosterna fusca*)—cause no little damage to the lawn. The latter is not satisfied with the intolerable annoyance he causes in the imago form by bumping against everything he sees, but already begins in the pupa stage to devour the roots of grasses and valuable plants, blighting everything his voracious mandibles seize upon for prey. Patches of dead and withered grasses proclaim his depredations, when the turf should be closely perforated with a metal rod to the depth of half a foot, pouring caustic lime into the openings, and resowing the surface a few days afterward.

The ant is fond of building his cities on the sward. These may be destroyed by perforating the hills and pouring in a solution of crude carbolic acid, composed of one pound of acid to two quarts of water. A gill of the liquid will suffice for an ant-hill. "Tobacco insecticide soap" is also efficacious. It is, moreover, excellent, when sufficiently diluted, for destroying ants where they have formed their hills in or about plants. We thus see that a fine, velvety sward, like very many other desirable things, has its price; and that, to carry out Loudon's apothegm, " The basis of all landscape-gardening is good breadth of grassy lawn," calls for forethought, pains, and unflagging perseverance.

After the lawn, the flower-border demands our attention. And here, especially, I repeat with emphasis the golden rule of the garden: That is worth doing well what is worth doing at all. Compare the sickly, starveling flowers, struggling for bare existence in beds choked with weeds, and baking in a crust of arid earth, with the luxuriant, well-grown plants which careful culture yields. In the one case, disappointment; in the other, constant increase of beauty. "But I am no gardener, and Primrose employs a professional," is the reason often assigned; the important fact being lost sight of, that back of the gardener and all other garden operations lies the fundamental principle of floriculture—*proper preparation of soil*. The parable of the sower has also its application to the garden.

A rich, friable loam is adapted to the requirements of the majority of border-flowers. Where the natural soil is stiff, clayey, or sandy, it is useless to expect satisfactory results, even with the most liberal manuring. Clay soils can only be rendered tractable by the addition of leaf-mold, sandy loam, and decomposed manure in sufficient quantity to render the soil free and elastic. Sandy soils should be treated with plenty of strong garden-loam, leaf-mold, and an abundance of old manure. It may be observed, in this connection, that a leaf-mold and compost-heap should form a part of the reserve garden. When leaf-mold is desired, it is often difficult and expensive to procure. The rakings of old leaves in autumn, and the leaves used for winter protection, left in a heap to decompose, will usually suffice to keep up a sufficient supply. Proper drainage secured, the flower-border should be composed of surface-soil to the depth of at least two feet. This will insure the roots a supply of moisture far below the surface. Treated thus at the beginning, the foundation will be permanent; and, beyond sometimes forking in an autumnal top-dressing, we have done with the question of soil. I am aware that it is often the custom in England, where climate and skill produce the highest results, to retrench and replant the flower-border every three or four years. This involves much labor, and disturbs numberless plants which do not like removal. It is far easier and better to separately lift or divide such plants as may have exhausted the soil, replanting them in fresh earth.

No plan of gardening involving an expensive

annual effort can be satisfactory, even to those of abundant means. It should be the effort, therefore, to plant subjects that will be permanent, and increase in beauty from year to year. If a plant proves unsatisfactory from any cause, cast it aside. If its color clashes with that of its immediate neighbors, shift it to some other position where it will not offend. It is almost impossible to plant a large collection without color discords. The various shades of red in juxtaposition are generally the most trying, and, often, effects can not be fully measured until flowering-time. In such cases it is best to immediately shift one or the other offender. If left until fall, even when a detailed garden memorandum is kept, the cause is apt to be mistaken or forgotten, to intrude itself another season.

With comparatively few exceptions, transplanting may be effected even during the hottest weather by soaking the soil, lifting the plant with a ball, and replacing in soil which has been thoroughly watered. In dry weather the soaking must be thorough and repeated, so that the subject may be lifted with a good-sized ball, and little or no root-disturbance. This operation is best performed in the evening. In hot, sunshiny weather the plant may be shaded for a few days until re-established.

Often plants crowd each other; too many species of similar habit occur side by side; hurried spring planting may place desirable subjects amid incongruous surroundings, and the symmetry of the flower-border become disturbed. Its outline, shading, and harmony of color are naturally seen to the greatest advantage at the flowering season, and it is then that transplanting may be most intelligently performed. Certain subjects, like lilies, daffodils, etc., must, of course, await their proper season for removal; and, where the subjects for shifting are numerous, cool, wet weather should be selected. I would not by any means appear to advocate summer transplanting, to the exclusion of spring and fall; but where the position of individual plants is immediately offensive, or where they are unduly crowding each other, summer transplanting is to be recommended.

It is always advisable to have a reserve flower-patch to draw from, where subjects may be obtained to replace those that may fail or prove unsatisfactory, for the purpose of exchange, or where masses of particular kinds are liable to be called for. Generally, a stock of desirable plants may be had by annual sowings and division. The seeds of some perennials germinate very slowly, and are often trying to raise. Much is to be gained with the majority by sowing as soon as the seeds are ripe, and wintering the slow-germinating kinds in a cold frame, pricking off when large enough, and planting out subsequently in their proper places. Not a few perennials spare us this trouble by sowing themselves; many bloom the first year where sown early; a large portion germinate slowly. In all cases, fresh seed insures the best results. Sow in light soil in shallow boxes, covering with a light layer of moss to retain moisture, and wintering in the cold frame such species as do not develop sufficiently to plant out in autumn. Perennial seeds one should not despair of until well on to the second year after sowing. Many of them are in the habit of lying dormant for a year. In England seed-pans are usually kept dark, being moved into the light as soon as the seeds are up. Lichens, which clog the surface of the soil, do not grow in the dark. Annuals germinate readily, and cause little trouble.

Another mode of propagation is by cuttings. These, taken from the plants just when growth begins, or after blooming, should be inserted in boxes or pots filled with a mixture of leaf-mold and sand, keeping them in a shaded frame until rooted; then pot singly in three-inch pots, planting them out finally the following spring where desired.

What flowers shall we plant, and how shall they be planted? This will depend largely on the space to be filled, and on other considerations. Many, who are accustomed to be absent during the summer, will plant spring bloomers almost exclusively—a mistake, for this means bare borders during midsummer. Where one has a rock-garden, some plants, that otherwise might find a place in the borders, will be kept apart in this more proper situation. Where there is a hardy fernery, ferns will naturally be excluded. There will also always exist a diversity of opinion regarding the merits of particular plants. Certain perfumes delightful to some are disagreeable to others; while, so long as people exist who can endure magenta passively, we may never hope to exile such nightmares as *Achillea rosea* from the border, or some of the shades of the Cineraria from the greenhouse. All hardy plants, desirable and beautiful themselves, which will thrive in the soil and position chosen, and which are not so small as to be lost in the border, may be used appropriately; these will be alluded to specifically, later on, in their order of flowering.

Experience will teach what not to plant better than volumes of instruction. Usually, subjects that sucker and throw out strong, creeping root-stalks are objectionable. Do not introduce rows in the borders; plants are not supposed to be on military review. Neither dot the ground at equal distances with the same plants often repeated; variety is the spice of the garden. Though the taller-growing species, as a rule, are best placed in the background, an occasional colony of large plants should be placed in the center, and some large individual specimens relieve the foreground. Massing, where too much space is not called for, is desirable, especially with plants of medium size; though attention must be paid to selection, or large bare spaces after blooming will obtrude. Where daffodils are largely grown, summer and autumn flowering subjects, like the columbines and Japanese anemones, should be placed in close proximity, to fill the void left when the bulbs die down in summer; or light-rooting flowers, like the lovely Iceland poppy and some of the finer small annuals, may be employed to take their place.

The great secret of successful gardening is continuity of bloom—a luxuriance of blossom from early spring to late autumn; so that, when one species has flowered, there will at once be something else to continue the blosssoming period without leaving unsightly gaps of bare ground. The necessity of placing plants intelligently will thus be readily apparent—the just apportioning of spring, summer, and autumn flowers with these several ends in view. Moderate shade is of advantage to many flowers, but this should never be obtained from trees planted in the border itself.

Plant permanently, mass boldly. Do not confine yourself to a few kinds when there is such a wealth to choose from—plants for sunshine and plants for shade, plants for color and plants for fragrance, plants for spring and plants for autumn, plants for flower and plants for form. Aim at individuality, to produce an ideal of your own. Many half-hardy plants in the accepted sense can be grown by simply protecting them with leaves over winter. Plant for permanency lilies, irises, roses, delphiniums, phloxes, spiræas, hemerocallis, narcissi, columbines, day lilies, herbaceous pæonias, bell-flowers, anemones, fraxinellas, perennial sunflowers, the great and lesser poppies, centaureas—the list is inexhaustible. Avoid coarse, weedy subjects, unless in special cases where habit may be compensated by bloom or special adaptation to situation ; these are usually best placed by themselves in the distance or the rear garden. Many an old-fashioned country garden can teach us much on the subject of selecting proper border flowers. The flower-border may be raised very slightly, to insure perfect drainage and to emphasize its contour, but never be so elevated as to cause over-dryness; elevated beds and borders are designed for plants which do not require much moisture. The skillful planter will not forget to place showy subjects with reference to their effect from the interior of the house, so that the beauty of the garden may be admired from within during inclement weather.

A garden may be rendered beautiful from early spring until late autumn with perennial flowers alone; but it may be rendered still more attractive by the judicious use of many of the finer annuals, biennials, and foliage plants as well. By the term "judicious" I mean not only a use of annuals of merit, but annuals properly placed; perennial flower-borders should consist in the main of perennial flowers. To cultivate hardy flowers it is not necessary to be an Asa Gray, though a knowledge of botany must always afford an ever-increasing satisfaction and pleasure. A love for flowers one must have; one can not be a Peter Bell in floriculture.

Finally, the garden syllabus may also be written on two tables of stone:

I. Whatever is worth growing at all is worth growing well.

II. Study soil and exposure, and cultivate no more space than can be maintained in perfect order.

III. Plant thickly; it is easier and more profitable to raise flowers than weeds.

IV. Avoid stiffness and exact balancing; garden vases and garden flowers need not necessarily be used in pairs.

V. A flower is essentially feminine, and demands attention as the price of its smiles.

VI. Let there be harmony and beauty of color. Magenta in any form is a discord that should never jar.

VII. In studying color-effects, do not overlook white as a foil; white is the lens of the garden's eye.

VIII. Think twice and then still think before placing a tree, shrub, or plant in position. Think thrice before removing a specimen tree.

IX. Grow an abundance of flowers for cutting; the bees and butterflies are not entitled to all the spoils.

X. Keep on good terms with your neighbor; you may wish a large garden-favor of him some day.

XI. Love a flower in advance, and plant something every year.

XII. Show me a well-ordered garden, and I will show you a genial home.

The Spring Wild Flowers.

Shall we be so forward to pluck the fruits of Nature and neglect her flowers? These are surely her finest influences. So may the season suggest the thoughts it is fitted to suggest. . . . Let me know what picture Nature is painting, what poetry she is writing, what ode composing now.—THOREAU.

III.

THE SPRING WILD FLOWERS.

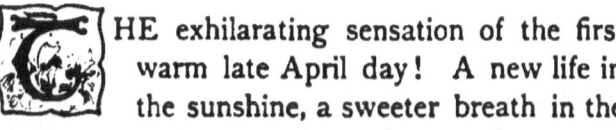HE exhilarating sensation of the first warm late April day! A new life in the sunshine, a sweeter breath in the south wind: the breath of green fields and re-animated woodlands; the fresh, unctuous smell of the soil! To it every living thing responds — the awaiting birds, the dry chrysalis, the imprisoned flowers. How merrily bluebird and meadow-lark ring out their welcome! With what a rush *Hepatica*, bloodroot, spring beauty, and dog-tooth violet burst through the mold! How all the wild, glad host of pulsating things seems eager to roll away the resurrection-stone!

I never see and feel the start of vegetation without recalling Rémy Belleau's sixteenth-century lyric on April, which still exhales the very essence of spring—a lyric unsurpassed by any I am acquainted with on a similar theme. To

April the French poet assigns a place exalted above all the other months: "*Avril, l'honneur et des bois et des mois!*" Unsurpassed in the original, the apostrophe is admirably rendered by Andrew Lang:

> April, pride of murmuring
>   Winds of spring,
> That beneath the winnowed air
> Trap with subtle nets and sweet
>   Flora's feet,
> Flora's feet, the fleet and fair. . . .

Nothing could be more truly descriptive of the mad hurrying into life of the spring flora than the spirit and *allegro* throughout the poem.

I think the first of inanimate wild life to pierce the ground is the well-known member of the aroids, the skunk-cabbage (*Symplocarpus fœtidus*). A rank, foul, noxious weed, "a noisome hermit of the marsh," it is usually considered—surely an unjust stricture. It has a clean, wholesome smell, a pungent, growing, out-of-doors smell, with no taint of corruption. Greuze would have admired its lovely greens, and, I doubt not, a poet will yet be born to praise its rugged precocity. I have planted it in the rear garden, on the edge of the copse, as a wild foliage-plant, just to watch its incurved horn and gigantic leaves expand. So long as we grow the crown imperial, we can well overlook the odor of the great green aroid which so boldly ushers in the spring.

The infinite shades of green which Nature has in her color-box! I say nothing of the marvelous greens of her twilight skies, or those of her streams and waters, but simply the greens of vegetation. There is another autumn of color in the spring foliage, so varied are the shadings of the buds and young leaves. Indeed, it is often difficult to tell where green begins or ceases, so interblended it is with reds and yellows. The different colors of the soil, too, what variety they present! there is almost a rainbow in the clays alone! I do not remember having noticed magenta in either foliage or soil. When Nature uses it in a flower, she is rather sparing, or gives it a proper foil of green to tone it down; its wild, barbaric effect she leaves to frescoes, florist's cinerarias, and Bahndur rugs.

Once started, the wild flowers succeed one another with astounding rapidity. The arbutus appears blushing almost beneath the snow, and so quickly is it followed by the many other early flowers that it becomes difficult to place them in their proper succession. A sheltered situation where the sun concentrates its warmth often calls out a species before its regular time, intercepting earlier species in less favored localities. Many of the flowers that we shall meet in the swamps and woods will be found worthy of a conspicuous place in the garden. Few realize the richness of our native flora. Comparatively few are familiar with its infinite grace and beauty in its chosen haunts. Fewer still appreciate how many of our wild flowers thrive under proper cultivation, or how much they add to the charm of the garden. Nature shows us the effect of liberal planting and bold massing. The woodlands hold no bare patches; each flower is quickly succeeded by another. The ground now glowing with the little spring beauty (*Claytonia Virginica*) will soon be painted with violets and *Trilliums;* and where the *Hepaticas* run riotously over the hill-side, ferns and flowering plants innumerable will take the place they have vacated. The *Hepatica* is one of the earliest flowers to extend an invitation to the woods. It grows on sandy hill-sides, frequents open glades, hides in shady hollows, and, like Montgomery's daisy, "blossoms everywhere." In color it varies from a lovely blue to pure white, shading to lavender and a soft flesh-tint. The spring beauty is scarcely less charming, and is even more prodigal in moist places. Not satisfied with one color, its flower-clusters also assume several hues--white, with shadings of rose, and penciled with deeper-colored veins. There is another form of spring beauty (*C. parviflora*), from Oregon, equally free blooming, which flowers later and spreads freely from seed. Besides these, I find five additional forms mentioned in the "Botanical Survey of the Fortieth Parallel."

Whoever has been in the woods in early spring has met the bloodroot (*Sanguinaria Canadensis*), with its white, star-shaped corolla, the delicately scented flowers preceding the large, kidney-shaped leaves. Its only fault is its ephemeral nature; you scarcely obtain a glimpse of it ere it is gone. It belongs to the poppyworts, nearly all of which are familiar with the Horatian refrain:

Vitæ summa brevis spem nos vetat inchoare longam.

I suppose many flowers, like many people, have their faults, if such they may be called. Even the arbutus, if born again, I think, would wish to appear with fresher leaves.

When violets of every kind have jeweled the fields and meadows, and crept into the swamps and woods, there comes a sudden fall of snow. The great white flakes everywhere strew the ground, clustering round the beech-

boles, flecking the hill-sides, dotting the slopes—
the chaste, pure triangles of the white wood-lily
(*Trillium grandiflorum*). Individual among
flowers, the *Trillium* is scentless—lovely enough
without perfume. To enjoy its full beauty, you
should come suddenly upon it in its wild-wood
home, or naturalize it with the bloodroot by the
hundreds, under trees or in shady spots in the
garden. It will hardly bear the shortest journey
after cutting. If you would have it in the
house, you should grow it in large potfuls,
treating it like the narcissus. The English pronounce it one of the most beautiful of hardy
plants, and I exchange it every year, with
friends in Cheshire and Kent, for *Horsfieldi*
daffodils. The purple variety (*T. erectum*) often
keeps it company. It is a jaunty flower at
home, but somehow appears out of place under
cultivation. *T. erythrocarpum* is a very pretty
species, fluttering a small white corolla with a
lively carmine eye. I found it swarming in the
Adirondacks with the large white and purple
varieties.

In "Les Fleurs de Pleine Terre," which I
opened by accident on page 1151, it is amusing
to read, under " *Trillium grandiflorum*," " The
Trilliums are curious rather than pretty plants,
and rather delicate, perhaps." To have the

Trillium thus characterized provokes a smile. A strange flower it certainly is—its leaves, calyx, and corolla a triangle. In the same volume I find the bloodroot described as "curious and pretty"—a distinction with a difference. The *Trillium* may be raised from seed—a much more tedious process than obtaining plants from the woods. It likes rich, deep leaf-soil and shade, requiring at least two years to become thoroughly established. Where *T. grandiflorum* is well grown, it often attains a height of nearly two feet. Not the least charm of this variety is its change to a soft rose-color—reversing the order of numerous flowers when they begin to fade. Indeed, variety and change of color in individual species is a characteristic of numerous spring flowers.

If the majority of our native violets have little odor, many of the very abundant species possess at least a faint scent, just enough to suggest an odor. The large-leaved *Viola cucullata*, and many of the tiny-flowered species, belong to this class. The bird's-foot violet (*V. pedata*) is less common than we would wish, more especially its variety *bicolor*, both species and variety having a rich, pansy-like fragrance, and velvety, pansy-like petals.

I do not think Bryant open to criticism for

ascribing fragrance to his yellow violet, blossoming

> Beside the snow-bank's edges cold.

The *Violas* are so associated with odor that it is difficult to think of any as entirely scentless. From the hosts of blue, purple, lavender, yellow, and white species that carpet the ground, and which, except the white *blanda*, are usually considered odorless, there certainly does arise a perceptible fragrance, perhaps best described by Bryant as a "faint perfume." Lorenzo de' Medici, a distinguished gardener and floriculturist himself, tells us in sonnet-form how the violet came blue. Originally white, Venus, seeking Adonis in the woods where it grew, stepped upon a thorn, which, piercing her foot, caused the purple drops to fall upon the flowers—

> Tingeing the luster of their native hue.

Shakespeare's violet was *V. odorata*, common in Europe and in many portions of Great Britain. "*Viola odorata* flowers all winter, but chiefly in March; the typical color is a deep purple-blue," Rev. Wolley Dod, of Cheshire, writes me; "it is not unlike indigo-dye, but in gardens there is every shade, down to pure white, the latter being, I think, the sweetest of all." The passage in which the violet figures most conspicuously, most beautifully, in literature, is too well known to be repeated. We can readily comprehend the comparison to Cytherea's breath ; but the reference to color—if reference to color was really intended—is less apparent on close analysis. Why, in the first place, should the lids of the goddess be singled out rather than the orbs themselves, which Shakespeare might have stamped indelibly a violet-blue ? Unfortunately, we have no data to fix the precise hue of Juno's eyelids, but we would naturally presuppose them to be dark. The old French abbé-philosopher, Brantôme, who, it must be conceded, is excellent if somewhat plain-spoken authority on all that appertains to the charms of lovely woman, specifies, in the " Vies des Dames Galantes," at the conclusion of his second discourse, " De la Veuë en Amour," that, among the thirty essentials which go to compose a supremely beautiful woman, there must of necessity be three black (*trois choses noires*)—the eyes, the eyebrows, and the eyelids :

Trois choses blanches : la peau, les dents et les mains.
Trois noires : les yeux, les sourcils et les paupières.
Trois rouges : les lèvres, les joues et les ongles.
Trois longues : le corps, les cheveux et les mains.
Trois courtes : les dents, les oreilles et les pieds.

Trois larges : le sein, le front et l'entre-sourcil.
Trois estroites : la bouche, la ceinture et l'entrée du pied.
Trois grosses : le bras, la cuisse et le gros de la jambe.
Trois deliées : les doigts, les cheveux et les lèvres.
Trois petites : les tétins, le nez et la teste.
Sont *trente* en tout.

Dark eyelids—a dark purple, rarely the typical violet hue—are a well-known mark of feminine beauty. Cleopatra's eyes must have flashed over them; and we know the fair Georgians of the East, who do not come by them naturally, use *k'hol* to produce the languorous charm they are supposed to impart. Still, this does not satisfactorily explain the Shakespearean analogy—

Violets *dim*, but *sweeter* than the lids of Juno's eyes.

To carry out the comparison of the poet, who distinctly qualifies the color as "dim," Brantôme's beauty-mark will scarcely apply in its literal sense. Possible allusion to fragrance is out of the question; it must, then, refer to some other sense—either to that of sight or feeling—the term sweeter being employed for lovely, or to denote softness to the touch. Let us, therefore, look deeper into the eye of woman. A kiss upon the eyelids—and for this we do not require

Gallic authority—is pronounced one of the sweetest things of life. This theory, then, may furnish the key to the passage; it is to the qualification "sweeter," in the sense of softer, not to the color-definition, that we must seek for its intended significance. On the other hand, if impetuous Jove kissed Juno, as there is every reason to suppose he did, we must conclude that he preferred roses to violets, and kissed her on the mouth, and not on the eyelids. Clearly, this is a subtle ruse of Shakespeare, all the more abstruse from its lovely imagery, and is only another case of "The Lady or the Tiger."

Passing from the "Winter's Tale" to the "Country Churchyard," the verse printed in two editions of Gray, and then expunged from the "Elegy," presents itself:

> There scattered oft, the earliest of the year,
>   By hands unseen, are showers of violets found;
> The redbreast loves to build and warble here,
>   And little footsteps lightly print the ground.

Why Gray should have canceled this exquisite stanza is inconceivable. It is the relief, the very flower of the ode—the one expression of lovingkindness and human sympathy to diffuse warmth and fragrance over the tomb.

Finally, before taking leave of the violet, I wonder if a resemblance of two poems, to which the spring flower's fragrance clings, has been noticed? I refer to Collins's ode "On Fidele supposed to be Dead," and Oliver Wendell Holmes's verses "Under the Violets." Both are pervaded by a pathos equally tender, the meter being alike, except the added fifth line of the latter. Though a similarity will be observed, consisting rather in meter, pathos, and sentiment than in any direct expression, it is not difficult to pronounce upon the comparative merits of the two poems. Viewed by posterity, assuredly Holmes's will be regarded as the richer, the more finished ode:

>   To fair Fidele's grassy tomb
>   Soft maids and village hinds shall bring
>   Each opening sweet of earliest bloom,
>   And rifle all the breathing spring.
>
>   .   .   .   .   .
>
>   The redbreast oft at evening hours
>   Shall kindly lend his little aid,
>   With hoary moss, and gathered flowers,
>   To deck the ground where thou art laid.
>   COLLINS.

>   For her the morning choir shall sing
>   Its matins from the branches high,
>   And every minstrel-voice of spring
>   That trills beneath the April sky,
>   Shall greet her with its earliest cry.
>
>   .   .   .   .   .

> At last the rootlets of the trees
> Shall find the prison where she lies,
> And bear the buried dust they seize
> In leaves and blossoms to the skies;
> So may the soul that warmed it rise.
>   HOLMES.

While the violets are yet in the hey-day of their beauty, there is no lack of other vernal flowers. The adder's-tongue (*Erythronium Americanum*), almost first to dart its sharp purple spathe through the ground, appears in legions. The warmth has brought out the brown spots upon the now clouded gray leaves. Presently will appear its nodding, tulip-scented yellow blossom, revolute in the sunshine. Singularly, the adder's-tongue has its two leaves of equal length, but one almost double the width of the other. I do not find this dog-tooth violet a satisfactory subject to naturalize; it has a ragged look out of its native quarters, and even there it is not always as free-flowering as we would wish. The robust variety, *E. grandiflorum*, and the large, white form, *E. giganteum*, from the Rocky Mountains, are far more beautiful. A variety named *E. Hendersonii*, with lilac flowers and a central purple blotch, edged with yellow, discovered very recently in Oregon, is said to be the finest of the genus,

Little later than the adder's-tongue comes the lung-wort (*Mertensia Virginica*), pretty in the blue and lilac shades of its drooping flowers, and almost equally beautiful in the rich, dark purple of the early leaves. The large blue flag (*Iris versicolor*), an inhabitant of wet places in woods, meadows, and along streams, is a handsome subject for naturalizing where it can obtain the necessary moisture. Soon the little Dutchman's breeches (*Dicentra cucularia*) will disclose its curious spurred flower, and the columbine (*Aquilegia Canadensis*) plume the rocks and enliven the dry places with its pendulous scarlet-yellow blossoms. The wild crane's-bill (*Geranium maculatum*) is usually found with the columbine, both being fond of places where the *Hepatica* has preceded them.

Jack-in-the-pulpit (*Arisæma triphyllum*) I regard as the coarsest of the aroids, not fit to associate with refined flowers; it looks more like a snake than a flower. The name—Jack-in-the-pulpit—sounds well, and doubtless has helped it to retain popular favor. Female botanizing classes pounce upon it as they would upon a pious young clergyman. But it is an arrant pretender, and should be called by its proper name, "Indian turnip," which befits it well. Let it pass for what it is worth, and pose not as a flower but as a carminative—its only virtue. "Parson-in-the-pulpit" they call the wild *Arum* in Great Britain. At Mentone, on the Riviera, the flowers of one of the aroids (*Arum arisarum*) are termed *Capuccini*, in allusion to the brown-cowled brethren of a neighboring cloister.

The bell-wort (*Uvularia grandiflora*), although far from being a monstrosity, is another plant that makes the most of its name. Unattractive, it is not hideous; neither is it brazen, like the Indian turnip. Instead of thrusting itself forward and demanding attention, it is rather graceful, hanging its head as if conscious of its dingy yellow. Its smaller sister, the dark, sessile-leaved bell-wort, is much prettier. On account of its creeping, deep-rooting rhizome, it should be avoided in the rock-garden, where it soon becomes troublesome.

In woods and on shaded hill-sides the rue anemone (*Thalictrum anemonoides*) is conspicuous—a dainty plant, with delicate foliage, and graceful white flowers assuming a blush tinge in some localities. It increases under culture, thriving both in shade and sunshine. A double form, which is in cultivation, is said to be even preferable to the common variety.

Now the shad-blow (*Amelanchier Canadensis*) has lighted its chandeliers and silvered the edges of the woods. It has seemingly a wild grace of its own, being seldom equally branched on all sides, but leaning its feathery sprays far over the woodland's edge. This is the case only where it is crowded ; for isolated trees, in nature or under cultivation, do not possess this habit, one of its charms in the woods. I have always envied those who can enjoy the white alder, or sweet pepper-bush (*Clethra alnifolia*), whose midsummer fragrance hangs like incense over the thickets where it grows. In August I should be willing to exchange it for the *Amelanchier*, only to regret it in May. The shad-blow has scarcely vanished ere the dogwood (*Cornus florida*) succeeds it as torch-bearer. A very much larger white flower, or, strictly speaking, involucre, it is scarcely more brilliant from a distance. It is far more distinct on close approach, and one would have to think twice to decide to which the preference should be accorded. I love the shad-blow, because it is first to appear ; and the dogwood, not only for its beautiful inflorescence, but for its brilliant red berries and glorious autumnal hues.

The dogwood is still in majestic bloom when the wild thorns add their tribute to the flowering pageant. Perhaps the thorn seems the showiest of the three, because it so often occurs as an isolated specimen. It has a pleasant way of surprising one, peering at you over precipitous banks, suddenly springing from some lonely hollow, or startling you by its snowy whiteness on some meadow or pasture. Have you wondered at the symmetry of many of these patriarchal pasture thorns?—the cattle have manipulated the pruning-shears. I think a gnarled old thorn, standing sentinel over a hill-pasture, the most picturesque of trees. For a century, perhaps, it has buffeted the wintry blasts, and escaped the shafts of the lightning, still to simulate perpetual youth in its perpetual bloom. The ground around it has been worn and trodden by countless hoofs; and on sweltering midsummer days the cattle ruminate, and lash their tails, beneath its woof of shade. It is the next thing to the shaded stream with white water-lily cups to keep it cool.

You look for the shad-blow with the snowy drifts of the *Trillium* and the running yellow flames of the marsh-marigold (*Caltha palustris*), that

> Shines like fire in swamps and hollows gray.

Hamerton calls the leaves of the water-ranunculus "the most beautiful of all greens in the world." Strange that he should have excluded the marsh-marigold, than whose glossy foliage nothing could be a lovelier, livelier green! A "gay, glabrous green, with glazed and brilliant yellow flowers," the publication that reviled the *Trillium* describes it very prettily and correctly. The *Caltha* is common to France also, and a Frenchman can always paint a French flower artistically, whether a wildling or a duplicate new rose. There exists a double variety, and also a white *Caltha*, a Californian species. The water-ranunculus (*R. aquatilis*) is a common American plant. It grows submerged, and floats a shabby little white flower on the surface of the water. Concerning the color of its foliage, which Hamerton extols, a botanical friend suggests that artists are apt to be enthusiastic about trifling differences which ordinary mortals do not notice. Of the tenants of the brooks and streams, the greens of the common marsh or water cress can scarcely be exceeded in beauty when swaying with every motion of the current. Do not imagine, because the *Caltha* grows so abundantly in the wet places, that it is easily cultivated, unless you possess the luxury of a bog-garden or a running stream for it to wade in, when you may naturalize it to your heart's content. One always wishes to transplant these water-loving flowers, they look so cool and seem to grow so easily. But they are born thirsty, and soon pine without their liquid nourishment. It will not suffice to give them a sponge-bath; they demand the bath-tub, and only luxuriate where their roots are forever drinking the moisture.

If you have a sharp eye and are acquainted with its haunts, you will see the large leaves of *Orchis spectabilis*, earliest of its family, pushing up to join the spring-tide pageant. The dwarf cornel has begun to prepare for its chase with the twin-flower and *Vaccinium* over the prostrate logs; while the bladder-fern and polypody crowd the stumps and bowlders, and the little *Cystopteris* is fast uncurling its interrogation-points.

One of our most beautiful wild flowers is the little fringed *Polygala* (*Polygala paucifolia*), its refined rose-red or purple flowers resembling a small sweet-pea. It rises from long, white subterranean runners, rambling over shady hill-sides with the goldthread and star-flower, and occasionally the fragile little oak-fern. Is there any blossom poised quite so airily above its whorl of lanceolate leaves as the star-flower; and could there be anything fresher than the dainty, shining foliage of the goldthread, that threads its leagues and leagues of golden runners through the cool, shadowy places of the woods? All these, with the dwarf cornel (*Cornus Canadensis*), itself a bold rambler and always fresh-looking, are charming when well established in the Alpine garden.

I should like to see a wild-woods garden placed in almost entire shade, and free from all rude draughts of air, composed exclusively of some of our native trailers and flowers, and a few of the miniature ferns. For the trailers, runners, and carpet plants, for instance, twin-flower, partridge-vine, goldthread, dwarf cornel, fringed *Polygala*, false Solomon's-seal, prince's-pine, ground-pine, and winter-green; with star-flowers, *Pyrolas*, bluets, and star-grass; and, for the small ferns, the common polypody, the oak and beech-ferns, the smaller *Cystopteris*, some of the dwarf spleenworts, and the hart's-tongue.

When Daffodils begin to peer.

In the flower-garden especial observance ought to be taken of the choicest roots of the Asian *Ranunculi*, Aulmoneys, tender *Narcissi*, and divers others of the like Tendernesse, and strangers to such Entertainments as our Northern Countries afford.—PHILOSOPHICAL TRANSACTIONS OF THE ROYAL SOCIETY, ARTICLE LXXXIX.

Devotion to Flora as a queen among us is as yet a living truth, and among or around the heart of all true gardeners there is woven a thread of twisted gold.—F. W. BURBIDGE.

IV.

WHEN DAFFODILS BEGIN TO PEER.

THE white-throated and white-crowned sparrows have lingered longer about the garden and the copse than usual before retiring to distant coverts. Thanks to unremitting warfare, my premises are comparatively clear of the English sparrow, so it is possible to hear the song-birds. Next to the incomparable music of the hermit-thrush, I think the major and minor of these two sparrows, who are almost always in each other's company, one of the most pleasing of all our bird-voices. They are more sociable than the hermit-thrush, who sings his hymn only in the most secluded woodlands; the latter has hurried past us this season, not making his customary pause on his return trip. The blackbirds have suddenly disappeared, after a brief dress-parade on the lawn. Over the distant lowlands I hear the vibrating warble of the meadow-lark; while high above the pastures float the mellow strains of the bobolink. The wood-thrushes are early and welcome arrivals. I wish they might remedy the disagreeable crack in their notes, which they seem to have caught from the grackle, the termination of the second bar frequently sounding like a snapped bowstring. Otherwise the notes would be very liquid, and, at a distance, might almost pass for those of the hermit. The Baltimore orioles have brought with them their orange-scarlet plumage, and still another new note which they will change from time to time. Year before last it was more sustained, and quite as plain as if one pronounced it, " Pretty, pretty bird!"

The same cat-bird—I am sure it is the self-same demon—has taken up his perch in the maple close to my sleeping-room, precisely as he has done for two years past. Nothing could be more delightful than his opening matin song, begun in a dulcet undertone, did I not know from experience his long-drawn *crescendo* and the frenzy of the *finale*—a perfect Hungarian "Czardas"! Pelting him with stones, a pile of which I keep within reach, stops him, as it does my morning nap. But he returns persistently to his chosen tree. I shall turn the garden-hose upon him some evening, and see if cold water possesses the virtue that the prohibitionists would have us believe.

Notwithstanding the caution I gave to spare the shears, the gardener ruined the beautiful *Forsythias* on the slope. If one needs an illustration of the cruelty of spring-pruning certain shrubs whose habit it is to flower on the old wood, he has but to trim a *Forsythia* into a rigid outline and compare it with one left untouched. All the airy grace of the golden sprays is fled. *Fortuneii* and *viridissima*, the former especially, are the best of the *Forsythias*, or golden-bells; *suspensa* looks ragged, even with close pruning.

If you commence early to plant magnolias, you may possibly succeed in obtaining one to solace your declining years. The money the nursery-men must make layering, budding, and grafting the acres of things they do, and then levying two or three dollars apiece on the wares they puff up in their trade-lists! All they do is to stick their things into the soil, and they take care of themselves. They must make thousands annually on magnolias alone; for there is no case on record of any one establishing a magnolia until at least three or four attempts. I find growers invariably recommend transplanting this tree, when in blossom, the last thing in spring—a cunning device to sign its death-warrant, so as to insure another sale the following year. *Magnolia Halleana*, or *stellata*, is beautiful on the lawns, with the *Forsythia* and the pink Chinese double-flowering plum (*Prunus triloba*). Every little while one feels like touching his hat to Japan, it has supplied us with so many valuable hardy shrubs and plants. *Conspicua* comes next to *Halleana*, a much larger plant and flower. *M. Lennei* is a dark, late-flowering variety which should not be overlooked. The scarce *M. purpurea*, while not nearly so robust, has a more refined and distinct flower than *Lennei*, of a very rich lake-color; the petals are narrower and more pointed than most magnolias.

In well-sheltered positions *M. macrophylla* will withstand even the severe climate of western New York, by protecting it for the first few years during winter—a fact worth remembering with regard to many deciduous and evergreen trees which are usually considered not perfectly hardy. This species would be worth growing for its magnificent leaves; when to these are added its gigantic white tulip-shaped blooms, it is incomparably the most tropical-looking of all our trees. To obtain its most striking effect it should be seen in a clump, the immense flowers being relatively few.

Here it is well to direct attention to the prevailing error of planting permanent subjects too closely, or too near walks and roadways. It should never be forgotten, when planting, that the small tree must grow, and eventually require space to develop. How often noble specimens, just when they are attaining their full beauty, must be removed, from this point having been lost sight of in the first instance!

Unfortunately, *conspicua* and *Lennei* are both somewhat tender; and of the large-flowered species, *Soulangeana* is on this account one of the most satisfactory for general cultivation. *M. Thomsoniana*, an American hybrid, a cross between the native *glauca* and *tripetala*, seems to have become lost of late years. Difficult to propagate, no doubt the nurserymen can not realize a sufficient dividend upon it, and so have discarded it. It is a valuable half-evergreen species, retaining much of the fragrance of its American parent.

Soon after the first magnolias the Japanese quinces appear, the most brilliant of ornamental shrubs. A single specimen of the scarlet variety will light up the largest lawn. There is a softer and equally beautiful shade in the varieties *umbellicata, aurantiaca,* and others; and also numerous lovely flesh-colored kinds.

The double-flowering white Japanese peaches have appeared with *Spiræas Thunbergii* and *prunifolia.* It is not because its blossom is whiter than the *Spiræas,* but because it so resembles the great flakes of the last flurry of snow, that the white peach seems the whitest of all flowering shrubs. The variety *versicolor plena* surprises one by its strange freak of producing variously white, red, and variegated flowers on the tree at the same time. It is nothing new to advise planting white-flowering trees and shrubs, with evergreens for a background; nevertheless, it is good advice always worth repeating.

The rose and red flowering peaches are likewise highly ornamental, and all the double-flowering cherries, notably the double white, may be placed in the same class. Most of the flowering crabs are beautiful. The blossom of the fragrant garland-flowering crab (*Pyrus malus coronaria odorata*) is not nearly as big as its name might imply, being a modest blush-flower borne in clusters, with the perfume of sweet violets. But while admiring this and many other ornamental flowering trees, let us not overlook the glorious inflorescence of the apple itself, a flower as tender in coloring and delicate in fragrance as the rarest exotic. "A rose when it blooms, the apple is a rose when it ripens," says John Burroughs, who has said about all that can be said on the apple in his own inimitable way. What a gardener he would have made had he followed Loudon as closely as he has Audubon! To properly enjoy Burroughs, he should be read in the author's pocket edition, published by David Douglas, Edinburgh. The burly, brown-cloth American volumes are too coarse a casket for the jewels they enshrine. The only possible objection to his locusts and wild honey is that they are sometimes too highly flavored with thyme from Mount Whitman.

The yellow-flowering or Missouri currant is in bloom. It deserves to be cultivated, if only for its odor. A shrub will scent a garden, and a bunch of it a hall; and its bouquet is as spicy as that of the yellow 'St. Péray wine, which I fancy it resembles, the favorite of Dumas *père*. The bees crowd around its yellow blossoms, and its honey should be worth its apothecary-weight in gold.

Herrick's *Julia* was born too soon. She missed *Horsfieldi* and many hundred others among the beautiful new English daffodils. But how much time she would have required to select a corsage-bouquet from the infinite number of nineteenth-century varieties, each one more bewitching than the other! I find three hundred kinds in Barr's catalogue alone, with scores of undiscovered ones running wild through the Pyrenees, and who knows how many more new hybrids to be heard from? Parkinson and Hale would have been beside themselves at the multitudinous forms and varieties. The daffodil is a flower for every one, and no spring garden is a garden in the full sense of the word without the grace and gayety it lends. Orchids are very well, yet they never seem to me to be a flower to excite special envy; we know they are beyond the reach of the masses, and that only a millionaire can grow them. Not so with the daffodil, which every one can enjoy in moderation, though a fine collection may be made a very expensive luxury as flowers go.*

Of all floral catalogues, a daffodil catalogue is the most exquisitely tantalizing. The further you read, the deeper the gold; and you are even met with

> Apples of gold in pictures of silver.

---

* The term daffodil I have used in its general sense. Specifically speaking, in many cases the term *Narcissus* would naturally be employed.

Daffodils running the entire gamut from yellow to white. Daffadillies with trumpets flanged, expanded, gashed, lobed, serrated, and recurved. Daffadowndillies with perianths twisted, dog-eared, stellated, reflexed, imbricated, channeled, and hooded. Then the multitudinous divisions and classes. Hoop-petticoat daffodils, single and dwarf trumpets, bicolor and shortened bicolor trumpets, white trumpeters, coffee-cups, tea-cups and tea-saucers, musk-scented and *Eucharis* daffodils, jonquil-scented and rush-leaved, goblet-shaped daffodils, *polyanthus* or *tazetta*, early and late poet's daffodils, jonquils, double daffodils, and how many more of the gilded host

To add to golden numbers golden numbers!

Lilies are tempting enough in the catalogues. But the lists finally come to an end, while the varieties of the daffodil are inexhaustible. The names, English and Latin, are so tempting, too, though these are nothing compared to the descriptions. To catch the daffodil-fever severely means either to break the tenth commandment or to be guilty of ruthless extravagance. You know there are swarms of varieties that will not succeed; but how are you to single them out without trying them? How artistically, how artfully devised some of the monographs are!

Sulphur hoop-petticoat daffodil (*Narcissus corbularia citrina*), for instance—as if the name were not enough to sell it—bears this description: "It is a bold and shapely flower of a soft sulphur tint, 'the color having a luminous quality, the flower being like a little lamp of pale-yellow light.'" Observe that two modern Parkinsons are called upon to describe it, so that, if one fails to hook the reader, the other will be sure to land him.

William Baylor Hartland, of Cork, Ireland, should be regarded as Herrick's and Wordsworth's successor. His illustrated "'Original' Little Book of Daffodils" is a very *epithalamium* of the flower of the poets. If we only had his climate and the Gulf Stream to help us raise his *Narcissi!* Like most flowers, the daffodil is thankful for careful culture. It dislikes manure, preferring good loam and a liberal sprinkling of sand. Climate, however, is everything with it. It likes to usher in the season gradually, not hurry it as our spring wild flowers do. Mild winters, gradual warmth, and abundance of moisture during the early season suit it best. For many kinds our springs are too sudden, and the transition from frozen ground to almost tropical suns is too rapid. In England, from February, when daffodils begin to flower, until May, the climate hesitates between winter and spring, and this is what daffodils seem to like. Nevertheless, even there some of the large trumpets go off with a kind of blight in masses after bad seasons. The flowering of the following year so depends upon the full development of the leaves that, if the weather suddenly becomes blazing and burns up the foliage, degeneracy is sure to result.

To the labors of the late Edward Leeds and William Backhouse we are indebted for many of the finest hybrid forms. Leeds was the prince of hybridizers, and was followed by Backhouse, who raised empress and emperor. Many of the hybrid *incomparabilis*, however, are so similar in form and coloring as to be perplexing and to uselessly extend the list of varieties. Of all these hybrids the *Nelsoni* are the finest and most distinct, with broad, snow-white perianths, and yellow cups usually suffused with orange on first opening. I was about to pass by the *Barri* varieties. But I find *B. conspicuus*, which has just opened, is almost another bicolor *poeticus*, also somewhat resembling one of the finest *Leedsi* forms, *aureo tinctus*.

Since writing the above, I find the reverse opinion maintained by Mr. Burbidge, one of the best authorities on *Narcissi*. "As a grower of nearly six hundred forms in a public garden," he says, "I know something of the variability of daffodils, and also of the taste of those who see them. Often and again will one visitor condemn a particular form which the very next will stop to admire. Some will even tell you that there is none or but little difference between John Horsfield and empress; whereas the differences are very marked in size, height, color of trumpet and of foliage, and in the date of blossoming. Taste is a shifting index, and there is room for all the varieties we now possess and more." Mr. Burbidge also imparts the information that those *Narcissi* possessing thick, fleshy, prong-like roots will grow anywhere, even in manured soils; but those having thin, short bunches of fine, wiry fibers will not do so, and must be grown in sand or gravel and pure fresh meadow-loam only.

Hybrids in the genus *Narcissus* are very readily made, and undoubtedly any species of the genus, under favorable conditions, will form a hybrid with any other species of it; and several of these kinds which are considered by botanists as species, seem to be hybrids; that is, they can be imitated by crossing two other species of the genus. The best-known instance of this is the so-called species *Narcissus incomparabilis*. A cross between *N. pseudo-narcis-*

sus and *N. poeticus* produces in some instances a daffodil which can not be distinguished from this; but the same cross may also produce results varying in the degree of each parent they contain, varying in the color, size of trumpet, and other particulars. These varieties are found wild on European mountains at elevations where *N. poeticus* and *N. pseudo-narcissus* flower simultaneously with the melting of the snow. It is this cross, made in gardens, that has produced all the Leeds hybrids. As for increase, some of the *incomparabilis* sorts multiply rapidly. Generally, orange Phœnix increases rapidly, but sulphur Phœnix never increases at all. The trumpets increase very irregularly; with me, *obvallaris* and the common *spurius* are perhaps the best growers of this section.

Among the bicolor trumpeters *Horsfieldi* and empress are incomparably king and queen. I confess I can perceive little difference between them aside from the foliage, except that the latter is a few days later to flower, and its trumpet stands out less boldly. Each exhales a rich magnolia-like odor; each flutters its pure white perianth and great golden corona over the luxuriant green foliage like some gorgeous butterfly, rather than a perfumed flower. Empress increases far more slowly than *Horsfieldi*. Its favorites claim for the former that it is better "set up," the perianth having more substance and the flower lasting longer.

The marked difference of the flowering period of these two and many other sorts is hardly apparent with us. Hot weather follows our cold weather so rapidly, that we almost lose sight of this distinction, and a great majority of the daffodils appear in blossom at nearly the same time. Emperor is certainly a grand variety, but infinitely larger in the English illustrations than in the American soil. Sir Watkin is scarcely as big as his name or his price would lead one to suppose. Nevertheless, he is assuredly the largest of the flat chalice-flowers or tea-cup section, and keeps on increasing from year to year. We must not expect to raise daffodils two to three feet high, as they can and do in England and Ireland, or grow them with trumpets large enough to serve the angel Gabriel.

Maximus (Hale's vase of beaten gold) I have been unable to manage. Neither can I grow the double *poeticus* successfully, after repeated trials with bulbs sent from England and Holland and procured here. It throws up strong flower-stalks, but they invariably come blind. I shall banish it to some neglected corner, where it will probably take better care of itself. *Ard-*

Righ, *nobilis*, *princeps*, and a form of single *Telamonius* are all distinct and desirable forms.

In a great vaseful of daffodils before me, *cernuus*, the drooping white *Narcissus*, is conspicuous, nodding lithely from its fluted stalk. Its sulphur perianth changing to white, and pale primrose tube, are heightened in their refined effect by its pendulous habit. It is a Spanish flower, and, as it can not wear a mantilla, it coquettishly hangs its lovely head. Smaller, but also beautiful, is *Circe*, one of the Leeds forms of the tea-cup section, with white perianth segments, and a cup changing from canary to white. The white daffodils generally possess a superior air of good breeding; they always seem dressed for the drawing-room. The yellow ones, even where they are superlatively handsome, look as if they preferred a romp or a game of tennis.

The Pyrenean *pallidus præcox* is invariably the first daffodil in the garden, closely succeeded by the distinct *obvallaris* or *Tenby;* the pale straw-color and cernuous habit of the one contrasting strongly with the vivid gold and large, wide-mouthed crown of the other. I have yet to see the daffodil which can compare with the intensity of its gold. "The causes of the singular and almost blinding intensity of the color," Hamerton explains from a painter's standpoint, "are a gradation from semi-transparent outward petals, which are positively greenish in themselves, and still more by transparence owing to green leaves around, to the depth of yellow in the womb of the flowers, where green influences are excluded, but yellow ones multiplied by the number of the petals. So in the heart the color is an intense orange cadmium, not dark, but most intense—a color that we remember all the year round." Hamerton says this in reference to Wordsworth's dance of the daffodils, and thus had *pseudo-narcissus*, or the common Lent lily, in mind, which has a pale perianth and rich yellow trumpet, and which is extremely difficult to cultivate in its native country.

Cynosure, another of the *Leedsii* hybrids, and Mary Anderson, single of the familiar orange Phœnix, are both strikingly beautiful. The former has a large primrose perianth changing to white, and an orange-scarlet cup; the latter, a silver perianth and a cup of lively orange-scarlet.

What with most flowers deteriorates from their beauty only increases the attractiveness of many of the daffodils, the fading perianth often adding a chastened beauty to the passing flower. Would that our pretty wives and sweethearts could all grow old so charmingly, or that woman might learn from the daffodil the art of always looking lovely things!

· The big trumpeters and chalice-flowers are not yet over before the *poeticus* and *polyanthus* groups and the jonquils appear. How cool the snow-white corolla of single *poeticus*, and how warm the rim of its dainty cup! And who that has ever scented it can forget its delicious aroma? The varieties of *poeticus* are many; the garden varieties, *recurvus*, *patellaris*, and *ornatus*, being finer than those collected wild. All of the *polyanthus*, or *tazettas*, are likewise deliciously odorous. The latter form pushes up so strongly in the fall, however, that it is apt to be injured by frost, and therefore the bulbs should be lifted after flowering and stored until late autumn. The big and little jonquils—and even here the variety is great—concentrate more odor in their little cups than any other form of narcissus. Of the double daffodils, *poeticus plenus* is too well known to be specified. With me, as has been previously observed, most of the buds come blind, the flowers forming inside the spathe, which becomes hermetically sealed, and soon dries up and dies. In England, where this species flowered very poorly the past season, a friend writes me that the same conditions prevailed, failure being attributed to the drought and cold winds of February and March, and something "going wrong" with it in May. The common double yellow is coarse compared with either orange or sulphur Phœnix. I can grow neither of these successfully. The latter runs out after the first year; the former gradually turns green—jealous, no doubt, of its thriving sisters in my neighbor's garden.

The hoop-petticoat narcissus of southern Europe I have yet to try out of doors, well protected in winter. It is of all the *Narcissi* the most individual, resembling an evening primrose enlarged and much lengthened.

The depth at which daffodils and lilies should be planted is a disputed question. In light soils it is well to err in planting too deep rather than too shallow; in stiff soils they should not be planted at all. Very many of the daffodils require to be placed in new soil every year or two; weak foliage and decreasing flowers indicate that they require a change. Transplanting, in either case, should be effected so soon as the leaves and stalks have died down, during the short space the bulbs are at rest. To secure the finest flowers, they should be cut in the full-bud stage, and allowed to expand in water within doors.

In England daffodils are taken up in July every year. James Walker, the largest grower near London, plants the bulbs in land that was manured for peas or early potatoes; a similar plan being adopted by the Dutch growers in their bulb-culture. Sea-sand is very genial to daffodils; the Scilly Islands soil consists of but little else. Constant replanting in deep, pure soil is the plan in England now, although five years ago growers were all manuring the soil for them. In Holland, *all* bulbs are lifted once a year. Fine crocuses, hyacinths, and tulips do not grow themselves. The soil in Holland is dark sea-sand or alluvium. Cow-manure is largely used for ordinary farm-crops, and after these have sweetened the soil it is dug over, two to four feet deep, and the bulbs are planted. Deep culture prevents their suffering from drought, and gives a clean, round bulb. To the Dutch should be awarded the prize for perfecting the bunch-flowering section, as to the English belongs the olive-crown for developing the grand trumpeters and the *incomparabilis* section.

For house-culture some of the *tazettas* are very effective, grown in the Chinese fashion, in water. Indeed, many of the *Narcissi*, which force readily, may be grown in this manner. In China *N. tazetta* is a favorite flower. The custom there is to place the bulbs in bowls of water with pebbles, the latter being employed for the roots to adhere to. But to produce Chinese effects we must have the Chinese narcissus, a splendid species, with immense, vigorous bulbs. The bulbs should be started in their receptacle with water about five weeks before they are wanted to flower, and placed in the dark until root-growth is made. They may then be moved to a sunny window, requiring no further care beyond keeping up the supply of water. They may even be grown in full light from the start. The Chinese *tazetta*, thus treated, throws up huge leaves, and stiff flower-stems two feet or more in height. There are two varieties, with single and double flowers, somewhat resembling in individual flowers *Grand Primo* and the double Roman *tazetta*, though of less substance and less highly perfumed.

Many of the lovely English hybrids we can not grow with success, owing to our rigorous climate. They are inversely like some of our wild flowers in England, which miss the frost and long season of rest, as some of the daffodils with us lack the genial climate they are accustomed to. Still, if many varieties refuse to become acclimated, there are very many others that are readily grown. Let us, then, follow the admirable precept of Délille :

Ce que votre terrain adopte avec plaisir,
Sachez le reconnoître, osez-vous en saisir.

I have been enjoying Délille in the old edition of eighteen volumes, copiously illustrated with quaint woodcuts. I found it in an old bookstall, and obtained it for a song. No wonder the late A. J. Downing was so fond of "Les Jardins," a French Georgic with nineteenth-century improvements! Sir Theodore Martin ought to do with this and "The Man of the Fields" what he has done with Horace and Heine; they are books that every gardener and lover of nature should be able to enjoy.

So many desirable forms of *Narcissi* may be had so cheaply, that almost any one can afford to grow some of the capricious varieties as biennials. With proper selection and intelligent cultivation, we may have in the daffodil a treasure-house of beauty, and, with this flower alone render any garden a field of the cloth of gold.

The Rock-Garden.

Pleasures which nowhere else were to be found,
And all Elysium in a plot of ground.
                        DRYDEN.

Imitez ce grand art, et des plants délicats
Nuancez le passage à de nouveaux climats.
Observez leurs couleurs, leurs formes, leurs penchans,
Leurs amours, leurs hymens.
                      DÉLILLE, L'HOMME DES CHAMPS

V.

THE ROCK-GARDEN.

 HEARD the *tremolo* of the toads for the first time, April 20th—later than usual. They are supposed to be silenced thrice by the cold—a rule I have generally found to be true. Though limited in compass, the toad possesses a musical voice, and only sounds it in warm weather. The orchestration of the small frogs, where each one tries to puff himself up as big as an ox, is emphatically a vernal tone, but it can not be termed musical. Their comical croakings always remind me of the peculiar noise made by boat-builders during the operation of calking. The huge, green bull-frog of the swamps, who is not heard until much later than his smaller brethren, has the merit of a powerful organ not entirely immelodious. In the distance, on hot summer evenings, his grand bassoon blends well with the

lighter and varied instrumentation of the lesser *reptilia*. His nocturne brings the plash of water and the scent of water-lilies nearer to me. It is a fluviatile expression, the fitting utterance of ponds and swamps. The cicada emphasizes no more tensely the heat of the midsummer noon, than the great batrachian the serenity of the summer night. His voice fits into the landscape like an audible shade—a sonorous emanation of coolness and departed day.

The trill of the toad is the prelude to spring, as the cricket's croon is the farewell to summer. How drowsily the chorus floats up from the lowlands—a summons to the early bees and flies to seek the precocious flowers! The blue scillas, the hepaticas, and the cowslips are swarming with the smaller bees and *muscæ*. Where do they come from in such swarms; and where do they all house themselves when the inevitable change of temperature puts a stop to cross-fertilization? A few warm days have done wonders toward starting delayed vegetation, each of the spring flowers apparently trying to outstrip the other. The pushing and striving for warmth and sunlight always seem to me among the most marvelous things of nature—the embryo seed, the rising stalk, the unfolding corolla, the perfect flower!

Scilla Siberica is perhaps the best of its class, although the comparatively new *Chionodoxa Luciliæ* is almost equally desirable for its lovely shade of blue. Of the other squills, the colors of *S. bifolia* vary much, some being far better than others; this species also contains a white variety. *S. Italica* and *S. amœna* are worthless. The later-flowering Spanish squills are large and coarse, but showy in shrubberies. These are of three colors—blue, white, and pink—sold under three names—*campanulata, patula*, and *nutans*. The difference in name does not always insure difference in flower. The best of all, certainly as regards color, is *S. Siberica*. From the chinks of the rocks the hepaticas glow with all shades of blue, purple, and rose, until they stop at nearly a pure white. The hepatica comes in the category of those flowers which the gardener neatly terms "very thankful." If you can not procure it readily from the woods, you should raise it from seed taken promiscuously from the different kinds, to procure new colors.

It is not strange that the British hold the primrose in such estimation that they have consecrated to it a "Primrose-Day"—April 19th—the anniversary of the death of Lord Beaconsfield, who wore a bunch of primroses in his button-hole whenever they were procurable. Hardy and floriferous, it is the richest of early spring flowers: from the palette of tints of the polyanthus, through the varied hues of the cowslip and common primrose to the "edged" and "powdery" *Auriculas*, the large, purple clusters of the Siberian *cortusoides*, and the fiery, opening eye of the Himalayan *P. rosea*. The Himalayan *P. denticulata* is a fine species, with bright mauve flowers on tall stalks. *P. Sikkimensis* is probably the most distinct of the Himalayan kinds, with lemon-colored and deliciously-scented trusses borne on lofty scapes. This must be raised from seed in pans or boxes; then, if planted out in shade in early autumn, the plants flower moderately well the following June. The second June they flower still finer, but after that they die, or deteriorate, and have to be replaced by fresh seedlings. It is one of the latest of its family to bloom. Nearly all the many varieties of the Japanese *P. Sieboldi* are charming, being perfectly hardy, unusually free-flowering, and remarkable for the size of trusses and flowers. A strain of English primrose, called Dean's high-colored hybrids, has produced some most tender and fascinating colors.

In many instances of primroses raised from seed, it is puzzling to know just where the polyanthus begins and the primrose leaves off—they seem to run into one another through hybridization. Our native primroses number but few species. *P. farinosa*, or bird's-eye primrose, also a native of Europe, is found in several localities. *P. Mistassinica*, a small, rose-colored species, rarely seen under cultivation, occurs in several Northern and Eastern States. The finest of indigenous species is *P. Parryi*, common in Nevada, the Rocky Mountains of Colorado, and the Uintahs, at an altitude of six thousand to ten thousand feet. This flowers from July to September, bearing fine rose-colored blossoms with yellow eyes, on tall stalks—a distinct and handsome hardy species. One should have a great bank of primroses placed in partial shade, to enjoy their fragrance and color *en masse*. And they should be raised from seed at least every other year, to keep up a supply of young plants, and to distribute among one's friends. But their most appropriate place is the Alpine garden, where they form dense cushions of bloom, and, with the daffodils, form a garden in themselves.

In English poetry the primrose shares an equal place with the violet and daffodil. It is referred to as the "lady of the springe," "winter's joyous epitaph," "merry spring-time's harbinger," "sweet infanta of the year," "the welcome news of sweet, returning spring," "the precious key of spring"; and most conspicuously by Shakespeare, who associates it with the daffodil and violet in the flowers let fall from Dis's wagon. Here, where it is comparatively scarce under cultivation, its beauties have only been sparingly sung by the poets, who nevertheless freely voice the praises of the snow-drop, crocus, and daffodil. Among our native flowers, the arbutus, violet, and gentian are freely singled out by the poets, and the azalea, bloodroot, hepatica, and cardinal-flower all come for their share of appreciation. I do not recall any poem on the spring beauty, the meadow-rue, the rue-anemone, or the moss-pink. Lowell is poet-laureate of the dandelion, and Emerson the bard of the rhodora. The wind-flower, or anemone, a well-known flower in American verse, would become a favorite, if only from Whittier's breezy lines:

> And violets and wind-flowers sway
> Against the throbbing heart of May.

Of all forms of cultivating flowers, rock-gardening is the most fascinating. Within a small space you may grow innumerable dainty plants, which would be swallowed up or would not thrive in the border—delicate Alpines, little creeping vines, cool mosses, rare orchids, and much of the minute and charming flora of the woods and mountains. Over this rock may trail the fragrant sprays of the twin-flower; here, at the base, a carpet of partridge-vine may be pierced by the wild-wood and meadow-lilies, and there a soldanella or Alpine gentian flash beside the fronds of an English fern. Then, its constant variety, and the inconceivable amount of plants it will contain! And how they develop and thrive among the rocks, where the roots have only to dive down to keep cool! I speak of the rock-garden as distinguished from the "rockery"—that embellishment to be found in company with the geranium-bed, surrounded by whitewashed stones; and iron stags or greyhounds standing guard over the growth of a hop-vine up a mutilated Norway spruce. With the "rockery" we are all familiar—that nightmare of bowlders, that earthquake of stones dumped out on to the hottest portion of the lawn, with a few spadefuls of soil scattered among them. Into this scant pasturage, where even a burdock would cry out for mercy, dainty plants are turned to graze. Fancy the rude shock to a glacier-pink or a Swiss harebell! The bowlder with a "pocket" is always at a premium, and within this parched receptacle, where nothing but *Sedum acre* or the common saxifrage could subsist, is placed a delicate Alpine. Of course, this is merely the death-warrant of the subject. Some tough and weedy species, that thrive on neglect, may survive the broiling ordeal. Usually only the rocks and *Sedums* remain, and the cultivation of Alpines is given up in disgust.

To grow Alpine plants successfully, it is necessary to understand the object of the rock-garden—its special adaptation to a very large class of beautiful plants, which find in it the root-moisture and natural surroundings they require. Many of these are too minute, many too fastidious, to be grown in any other way. The novelty, the delightful variety and charm which the rock-garden lends to the cultivation of flowers can scarcely be overestimated. From the very requirements of most Alpine plants, which love to run deeply into the soil in search of moisture, it is self-evident that there should be no unfilled spaces left between the base and surface. The rocks should be firmly imbedded in the soil, with sufficient space left between them for root-development of the plants. While the hideous chaos of stones of the average "rockery" can not be too severely condemned, half-buried bowlders, showing here and there their weather-beaten sides, have a picturesque look, especially when the flowering season is over.

The form of the rock-garden will depend largely on the character of the surroundings. Nothing can be more beautiful than a rock-garden at the base of a declivity, with the center, perhaps, forming a natural grotto half smothered with trailers and ferns. A rigid wall of rock will be avoided, while a round or even an oval mound is less pleasing than a form of somewhat irregular outline. Whatever form may be chosen, the rock-work should be constructed with a view of growing Alpine plants, and subordinating geological effects.

The soil is a matter of prime importance. Often, "potting-earth," as it is termed, is used, which becomes stiff and cakes badly during hot weather. For the majority of rock-plants a sandy loam proves most suitable. In some portions leaf-mold should be freely mixed with the soil, to meet the requirements of certain species; while peat-loving subjects will naturally be provided with the soil they prefer. A top-dressing of fine old leaf-mold and fresh loam every autumn will prove of advantage both in supplying the waste of soil from washings, and in serving as a fertilizer. I do not think the stress laid upon an easterly exposure, in England and on the Continent, applies here. The main points with us are shade and protection from draughts.

Spring subjects have mostly flowered before the trees are in full leaf; and, with our blazing summer suns, overshading through foliage will seldom occur. A few hours' sunshine during the day is sufficient for most plants which blossom after the latter part of May. The rock-garden is never appropriate in the center of a lawn. It is a dainty form of gardening, which should be enshrined by itself, rather than have its loveliness thrust upon one.

A rock-garden in a glade of a wood would be charming. This would afford abundant shade and moisture for the shadow-loving plants and diminutive ferns, as well as shelter from rude draughts, notwithstanding the belief, which most of us had when we were children, that it was the trees that made the wind.

Wherever it may be situated, it should be readily accessible to the garden-hose. I find a very fine dust-spray, which may be pinned into the ground and shifted from one point to another, the best means of watering. A coarse spray washes away the earth and is rude to the flowers. With sufficient moisture in summer and protection during winter, many species which are pronounced not hardy, or not to be acclimated, may be grown successfully. Oak and beech leaves covered lightly with evergreen boughs form the best means of protection. These should not be used until the ground is frozen, or plants may damp off, and mice harbor and cause destruction under the leaves.

Generally speaking, more especially where the space is limited, all plants with running, fast-spreading root-stalks should be avoided. Some of the harebells, for instance, desirable as they otherwise would be, are objectionable on this account. They must be hemmed in or have sufficient space, otherwise they encroach upon and soon smother their delicate neighbors. Some free-seeding plants are also to be guarded against. The *Sedum*, in many of its forms, is a pest, and with very few exceptions should never be introduced among rare and beautiful plants. I know of a rock-garden, admirably constructed at great cost, which had to be virtually torn apart to get rid of the *Sedum*.

The way really to enjoy the cultivation of Alpine plants is to build a new rock-garden every year, says Rev. Wolley Dod, one of England's most distinguished plant-culturists and botanists. I have been content with two thus far, and, so great is the enjoyment they afford, I shall supplement them with a fern rock-garden, for the smaller and more delicate ferns.

When referring to the toad, I omitted to

state that he is a treasure among flowers. He has a jewel in his tongue as well as his eye, and is better than whale-oil soap as an insect-exterminator. One would think his unwieldy presence must necessarily be destructive to fragile plants, yet his nocturnal hoppings leave no trace of injury to the most delicate flowers. How many gnats and flies and borers and aphides he snaps up with his sphinx-like tongue during the day, from behind the cool rock where he appears to be dozing, Gilbert White, I believe, has never computed. Richard Jefferies speaks of a strawberry-patch, the constant resource of all creeping things, where one toad always resided, and often two, and, as you gathered a ripe strawberry, you might catch sight of his black eye watching you take the fruit he had saved for you. The toad takes excellent care of the insects, but, unfortunately, can not manage the snails, which, unless carefully watched, are sometimes quite destructive to the tender leaves of certain plants.

Since the scillas, hepaticas, and spring-beauty have faded, another colony of flowers has appeared. The primrose yet remains, with tufts of later-flowering polyanthus and troops of merry-eyed auriculas. *Saxifraga cordifolia* and its varieties have thrust out their large trusses of rosy blossoms above their glossy leaves; and *S. peltata*, the gigantic species of the Sierra Nevadas, has sent up its tall stalks crowned with corymbs of pale-pink flowers, which appear before the huge, shield-like leaves. Two varieties of this species occur, one found at an elevation of six thousand to seven thousand feet, and the other growing in and along streams through the lower and warmer portions of California. The former is evidently much hardier and also more effective, its leaves in its native habitat often attaining a diameter of from three to four feet. *S. longifolia*, of the Pyrenees, is difficult to establish, but its near relation, *S. cotyledon*, which John Addington Symonds singles out as the finest of all the plants of the Alps, forms fine rosettes, although it has as yet refused to bloom for me.

The jonquils, *Trillium grandiflorum*, the rue-anemones, the tiarella, the purple and white *Phlox subulata*, the white *Erythronium* and *Trillium erythrocarpum*, are all in holiday attire. If we had not *Narcissus poeticus*, the latter might almost take its place, with its swan-white corolla and pheasant's eye. The rosy umbels of the garland-flower (*Daphne cneorum*) exhale such a delicious, penetrating perfume, that one is loath to leave it. Its opening crimson buds always tell me pleasant weather has come to stay. A native of the European mountain-ranges, it is one of the jewels of the rock-garden. But it is apt to prove capricious, and suddenly disappoint one by being winter-killed. Peat is usually prescribed for it. The finest specimens I have ever seen grew almost neglected, in rather poor, sandy soil, half-hidden by quack-grass. Gardeners should keep a memorandum, to strike a potful of cuttings every June, taken from as near the root of the plant as possible; cuttings grow slower, but make better plants than layers. *D. rupestris*, allied to *cneorum*, and the white *blagyana*, I have vainly attempted to establish. The former is undoubtedly hardy with winter protection, a microscopic plant having withstood two winters, and then dying off in summer.

The English nursery-men should be prosecuted for plant-infanticide. The miserable little sticks they send out are most of them too feeble to withstand a short journey, and, even with greenhouse coddling, are too weak and precocious to revive. The charges are certainly not at fault, for these would warrant adult plants instead of weaklings. Perhaps this stricture should not be confined to England, but apply equally to the Continent and America.

Of plants that grow in low-spreading masses several species of the *Phlox*, a genus exclusively

North American, are most desirable. *P. subulata*, or moss-pink, the little evergreen with lavender-colored flowers, together with the white and many other varieties, are all charming subjects. How gracefully, too, the moss-pink drapes a grave, paying its lovely but voiceless tribute to the departed! *P. procumbens* succeeds *subulata*, but neither its color nor its habit is as pleasing. *P. amœna*, with lighter-colored purple flowers and of dwarfer habit, is preferable to the latter. Prettier than either of these is a much larger growing species, *P. divaricata*, whose profusion of bluish or lilac flowers, on stems a foot high, perfume the places where it grows. Under cultivation, it increases rapidly in full sunshine. Growing near it, in a rich wood, I found, the other day, a colony of *Viola rostrata*, one of our most beautiful species, rare in this vicinity. It has a long, slender spur, the four lavender petals beautifully stained, and penciled with dark purple. The flower is of good size, and its hue might almost correspond to the "lids of Juno's eyes."

The white-umbeled, evergreen, sand-myrtle (*Leiophyllum buxifolium*) is in bloom, together with the yellow *Polygala lutea*, and the little yellow heath-like *Hudsonia tomentosa* of the New Jersey pine-barrens. There are very many easier things to grow; they demand a partially shaded position, and peat freely sprinkled with silver sand.

A host of Iceland poppies (*Papaver nudicaule*) has been called forth by the spring sunshine. They are, of all familiar poppyworts, the most beautiful, gracefully poised on tall scapes that nod and toss and flutter with every passing breeze. It is scarcely of these that Burns says:

> Pleasures are like poppies spread;
> You seize the flower, the bloom is shed.

Or Keats:

> At a touch, sweet pleasure melteth,
> Like to poppies when rain pelteth.

They are less fugacious than most of their widespread family, and there is always a fresh blossom to supply the one which has passed. The foliage is more delicate than that of any other species I am acquainted with, unless it be its little relative, the Alpine poppy (*P. Alpinum*). *Meconopsis Cambrica*, the Welsh wildling, somewhat resembles it, though it is coarser, more fugitive, and not nearly so floriferous. This does best in damp, sandy soil near water. As it is apt to die off the second year on dry soils, it is well to raise it from seed, which germinates readily. *Meconopsis Nepalensis*, the finest of the Himalayan species, I have three times failed to raise from seed; it is said to be a most capricious plant—either the seed is nearly always bad, or conditions are not favorable for germination more than once in two or three years.

Like all of its tribe, the Iceland poppy revels in sunshine, thriving best in sandy soil. All its forms are delicately beautiful, the yellow, white, orange-scarlet, and, rarest of all, a color I can only describe by comparing it to the plumage of the scarlet tanager. This is the only one of its species I know of which has a pleasant perfume. It is easily raised, and seed should be sown out of doors in August, or plants left to seed themselves. Occasionally among seedlings a semi-double form will occur, and also a very beautiful dwarf form, more frequently white than yellow, with short, stiff stems often bearing fifteen to twenty flower-cups within a diameter of five inches. A cream-colored semi-double form, with larger flowers than the type, is also very beautiful. I sow seeds of the white and orange-scarlet forms only; but of the latter the greater part come yellow. Though perfectly hardy, it is well to treat it as an annual, and thus always keep up a good supply of young plants to fill spaces made vacant by the daffodils when they die down, or to group freely in the borders. *P. umbrosum*, a hardy annual from the Caucasus, is larger, and not quite so neat in habit, yet strikingly beautiful with its dark-red petals blotched with black. *P. Hookeri* is another handsome annual recently introduced, extremely variable in the color of its brilliant flowers.

Gentiana acaulis gives us one of the most indelible blues of spring, a lovely, large, urn-shaped blossom clinging closely to the leathery leaves. An Alpine and Pyrenean plant, it is perfectly hardy and not difficult to cultivate. It is larger and more robust than its still prettier and near relative, *G. verna*, which opens its blue stars about a week later. This does best in a slightly shaded and well-drained position, and when abundantly supplied with water during midsummer. I may call it the sapphire of the rock-garden, as its exquisite blue flower is termed the gem of the mountain-pastures of southern Europe and Asia. Much later to appear is our own fringed gentian (*G. crinita*), mirroring the blue October skies, and exceptional for the four fringed lobes of its corolla. *G. Andrewsii*, also a native, has its deep purple-blue flowers striped within with whitish folds. You are fortunate if you can transplant the fringed gentian successfully; it is like the arbutus, and pines away from its home. All the gentians are beautiful and worthy of special culture; all, however, are difficult to raise from seed.

A classic flower, for it occurs in Greece and along the Mediterranean, is the scarlet wind-flower (*Anemone fulgens*). Its early flowering habit causes it to start so soon that, while unquestionably hardy with protection, it simply throws up its leaves without blossoming. In its own country it comes up in autumn, but the winters are so mild it does not suffer. It should be treated like the *tazzetta Narcissus*, and its tubers stored until November; a red wind-flower is so unusual a departure from the type that one can afford to bestow upon it special pains. *A. pulsatilla*, the European pasque-flower, distinguished for its large, solitary, violet-purple flowers, succeeds in well-drained limestone soil. The double of the common native wind-flower (*A. nemorosa*), discovered a few years since in Connecticut, is said to be a valuable variety, lasting much longer in bloom than the type. The snowdrop wind-flower (*A. sylvestris*), of Siberia and central Europe, is a lovely species, bearing medium-sized white flowers and blossoming in June, not unlike a small white Japanese anemone. *A. palmata*, *Alpina*, and *blanda* are all tender species, and so difficult to manage in England that it is scarcely worth while to attempt them. The anemone is poetically named from *anemos*, the wind, on account of the exposed places where it blows.

Without doubt *Iris reticulata* is the most beautiful of its tribe for the Alpine garden. Its early flowering habit, the beauty of its blossom, and pronounced violet odor, all render it exceptionally valuable. It blossoms well with me the first year, only to serve me like some of the daffodils and *auratum* lilies the second; a different soil, possibly, might tell a different story. No fault can be found with the common little *I. pumila*, likewise very early, and a species which increases rapidly. *I. cristata*, a very dwarf native species, produces large, handsome lavender flowers, blossoming almost on the ground from its creeping rhizomes.

All the *Iberis* are charming evergreen rock-plants, the coolest-looking of the spreading spring flowers. There can be scarcely anything more beautiful to cushion or overhang a ledge of rock than any of the forms of this hardy mountaineer. The varieties *corifolia* and *correæfolia* should not be confounded, for both are needed; the latter blossoming when the former has nearly passed. There is a blush-tinge to the large-flowered *Gibraltarica*, otherwise similar to the common *sempervirens*, though not so hardy. *I. tenoriana*, with purplish-white flowers, and *I. jucunda*, with small pink blossoms, also deserve a place. Desirable among white flowers is the hardy little Alpine catchfly (*Silene alpestris*), and the smaller *Tunica saxifraga*, that blossoms all summer. If you wish sheets of blue in June, *Veronica verbenacea* should not be overlooked, a pretty lavender-blue, and *V. rupestris*, a smaller, deeper-colored, and more compact variety. *V. pumila* is loosely habited and inferior to either of these. The diminutive *V. repens* is a valuable carpet-plant. It is the first of its tribe to appear, almost smothered with small pale lavender blossoms in early spring.

Of native wildlings, false Solomon's-seal (*Smilacina bifolia*) is easily naturalized in shade. The little yellow star-grass (*Hypoxis erecta*) will grow almost anywhere. Among trailing plants proper, there are none which exhale such a flavor of the woods as the twin-flower (*Linnæa borealis*), a favorite of Linnæus, and named in honor of the great botanist. It is not at all difficult to establish, as might be supposed, growing in sunshine, and luxuriating in light, moist soil and deep shadow.

The partridge-vine (*Mitchella repens*) is readily established, and is not over-particular as to a sun-umbrella. The partridge or ruffed grouse are fond of its sweet fruit, and hence the common name. There can be no prettier carpet-plant; when well established, it forms a thick mat of dark-green leaves covered with lilac-scented white flowers in June, and studded with brilliant scarlet berries in autumn. It is easily transplanted. Where it can not be had in large clumps, it should be gathered in preference from dry, sunny positions, and planted closely together, with a layer of chopped sphagnum on the ground between and all about it, being careful not to cover it. Where the space is ample, the false miter-wort (*Tiarella cordifolia*), also prettily termed foam-flower, may be used to advantage. A trailing plant, it is a vigorous grower, with large, shining, cordate leaves, and graceful racemes of white flowers in May.

The common winter-green, like the common polypody, generally prefers nature for a gardener. Even on dry hummocks where it occurs wild, it draws an element that it does not seem to find with artificial surroundings. I think there is much in the heavy condensation at night in and near woods and streams which explains the deterioration of numerous wild plants under cultivation; it is not always merely a question of soil, shade, or exposure. Many wild trailing plants succeed better when grown in large masses, doubtless because they thus retain the moisture longer. The winter-green, nevertheless, will do fairly well in shade, tightly packed in a mixture of old leaf-mold and loam. The goldthread (*Coptis trifolia*) is one of the finest of all small carpet-plants, and is easily naturalized in leaf-mold and partial shade.

Vaccinium macrocarpon, the common cranberry, is a fleet runner over the sphagnum, and bears transplanting even in sandy soil, where it forms a neat carpet, but not nearly so dense or of so thick a pile as the partridge-vine. With the *Mitchella*, *Coptis*, and *Linnæa* very many dainty native wild flowers may be associated, such as false Solomon's-seal, *Pyrola elliptica* and *rotundifolia*, wood-anemones, star-flowers, false violet, star-grass, and others. The little oak-fern and common polypody look pretty springing from the dark undergrowth. But the twin-flower, partridge-vine, and goldthread are so charming themselves that, in some places at least, the carpet should be formed of them alone.

Many of our native orchids are among the

most beautiful of plants for the shady portion of the Alpine garden. The showy orchis (*Orchis spectabilis*), the earliest of the *Orchidaceæ*, thrives under cultivation. The yellow lady's-slippers (*Cypripedium pubescens* and *parviflorum*) will do in the open border, but they never look appropriate, and the blossoms never attain the size or last as long as they do cultivated in shade. I have found both in nature, however, where the shade had been cut down, with thrifty stalks and well-formed roots. Indeed, the habitat of these two lady's-slippers varies extremely, both occurring (the large *pubescens* particularly) on dry, sandy banks and low, swampy woods; in marshy places the plants attain a far larger size and remain much longer in blossom. The showy or pink lady's-slipper (*C. spectabile*) is likewise easily grown when its natural surroundings are imitated; it is the showiest of all terrestrial orchids, and among the most distinct and beautiful of hardy plants. I find this does better, when transplanted, if the new shoot is cut out of the old wig of roots below it, the old roots seeming to encumber the plant. *C. acaule*, the stemless lady's-slipper, is a very handsome variety, erroneously thought to be almost impossible to establish. I find its purple flower sometimes in dry places, but commonly in damp woods. *C. arietinum*, the ram's-head lady's-slipper, a rare form, is easily cultivated in moist garden-soil with partial shade.

Of the *Habenarias*, *H. fimbriata*, the great fringed orchis which, with *psychodes*, is found in wet, rich leaf-mold, is not difficult to cultivate. They are both of marked beauty, the tall, brilliant purple spike of the former being a very conspicuous object in the woods. *H. ciliaris*, the yellow fringed orchis, is difficult to manage. I have been surprised to be most successful with the most delicate, *H. blephariglottis*. This is, I think, the loveliest of the *Habenarias*, attaining a height of from one to one and a half feet, with a spike of white-fringed, deliciously odorous flowers lasting long in bloom. Its habitat is cool sphagnum swamps, the plants springing from the clear moss, and never being at all connected with the soil. The white-fringed orchis should be planted in leaf-mold, with a ball of sphagnum about the roots, in full or nearly entire shade. *Arethusa bulbosa*, also a lover of wet places, and one of our most beautiful species, may be cultivated with success if good plants are secured to start with. *Spiranthes cernua*, or ladies-tresses, and *S. gracilis*, are neither of them difficult to manage in partial shade and sandy loam, and should be cultivated for their pretty, late-appearing flowers. Removal of most orchids may be made while the plants are in flower, and thus most easily found, by lifting them with a ball; great care must be exercised at any period, however, that the fleshy tubers sustain no injury.

Of British species *O. maculata* is the most satisfactory, the others being capricious, or finding something unconformable in our climate. The dark-purple blotches on the leaves of *maculata* are striking; and while the plant grows less strongly than at home, it nevertheless does well, its flower resembling a smaller *fimbriata*, but more variable in its shades. The British marsh orchis (*O. latifolia*) is one of the finest of the genus, bearing large purplish-pink flowers on a long raceme; it is always capricious and difficult to manage in its own country. The Spanish *Orchis foliosa*, which is not unlike *latifolia*, has wintered for three seasons with me, though as it does here it is inferior to either of our own fine purple *Habenarias*.

As to orchid culture, very few of the terrestrial species can be grown in sun with that degree of success which partial shade will give in skillful hands. The use of carpet-plants is often of benefit to the more delicate species, serving to keep the soil cool, and retaining the moisture about them; a few pieces of stone buried around them will answer a similar purpose.

Among suitable rock-plants which should not be forgotten are *Adonis vernalis* (the graceful rock-cress), the finer cinquefoils, many of the *Silenes, Saponaria ocymoides, Lotus corniculatus, Genista saggitalis*, the *Dodecatheons*, the *Alyssums*, the *Androsaces*, the Alpine *Dianthus*, and such of the Alpine harebells as do not spread too much at the root. The species and varieties specified in this, and alluded to in other chapters, are a few of many desirable plants suitable for the rock-garden.

There are hosts of others I am not familiar with that I have not enumerated; there are many that have not been alluded to because they are objectionable either on account of creeping root-stalks, bad colors, or other reasons; there still remain many tender or capricious subjects it is difficult to manage in our trying climate. But each one should try for himself plants which he thinks desirable, and thus ascertain their adaptability to soil and climate. I am informed, for instance, that *Onosma taurica*, one of the finest of Alpine plants, that is very difficult to manage in England and that has failed with me, is successfully grown in Boston. I might say the same of many other subjects that succeed in certain localities and fail in others. Capricious plants, however, should not be given up at the first failure. The old apothegm, "If at first you don't succeed," is especially applicable to many subjects of the garden.

The Summer Flowers.

Let us be always out of doors among trees and grass and rain and wind and sun. Let us get out of these in-door, narrow, modern days, whose twelve hours somehow have become shortened, into the sunlight and the pure wind. A something that the ancients called divine can be found and felt there still.
RICHARD JEFFERIES, THE AMATEUR POACHER.

VI.

THE SUMMER FLOWERS.

THE procession of summer flowers begins to form the latter part of May, and by the second week of June is well started on its march. A late or an early spring, a dry or a wet May, makes little difference with the state of vegetation on the first of the summer months. By that time the equilibrium is always reached, and Nature's balance-wheel is found revolving at its accustomed pace. Not until the advent of summer do the brilliant large flowers appear; the spring flora is smaller, more delicate, and generally more ephemeral. You must stoop down for the spring flowers; the summer flowers reach up to you. The procession formed in May and augmented in June moves steadily through July, when wild lilies blaze and tall *Habenarias* lift their purple spires; it moves onward during August over stubbles gay with vervains and willow-herb, and meadows fragrant with trumpet-weed; it files more slowly in September along streams flaming with cardinal-flowers and lanes lighted by golden-rod; until it halts and breaks ranks in late October, crowned with aster and everlasting, and strewed with painted maple-leaves. Do we half appreciate these summer days? We long for them in winter, and wish the months were weeks, to bring them nearer to us. Let us enjoy them when they come; let us get nearer to this joyous life of nature, and join in the procession of the flowers.

You would know by the scent of the lilacs that summer was here. How fragrant the censer of June! how profuse with the scent of blossoming vegetation!—odors not alone from myriads of plants, but breathing from orchards, hedges, and thickets, rising from woods and hill-sides, blown from far meadows and pastures. What an exhalation of millions of opening petals, mingled with the scent of green growing things! It seems as if Nature could not do enough when her appointed time arrives; as if there were no end to her prodigality of bloom and song and color and sunshine: birds singing amid the orchard-blossoms, bees plunging into the flower-cups, meadows smothered with buttercups, swamps golden with marsh-marigolds, woods aflame with honeysuckles, fields crimson with clover—bird-song, insect-hum, and flower-blossom on every side!

Among the large flowers of the garden, the germanica section of the irises is first to appear. To recommend any special varieties would be superfluous; they are so numerous, and are nearly all so beautiful. Easily grown, thriving in light soil and sunshine, we rarely see enough of them. This would not be the case if people would take the trouble to divide large plants, and thus not only obtain them more abundantly for another year, but increase the size of the flowers. The great bearded iris is one of the most effective border plants; the cut flowers are also beautiful when arranged with their sword-shaped foliage. The *Kæmpferi*, or Japanese section, is advancing, while the bearded iris is in bloom. Of these the varieties and colors are also innumerable; and, while more rarely seen, it is likewise one of the finest of perennials. Naturally a water-plant, it should receive abundance of moisture to acquire its full development. Where possible, it should be grown as a bog-plant. I should like to see it in company with the royal fern, sunk deeply in the mire. Where the space of the rock-garden will allow the use of large subjects, the Japanese iris may be appropriately employed. This species is so slow to advance, that its fine foliage retains its freshness for a very long period. The same observation will apply to the use of *Hemerocallis flava* in the rock-garden. The English and Spanish sections are much smaller species than either of the foregoing. Both have wonderful colors in blue, bronze, and gold, but are not to be compared with those above mentioned as border plants. *I. Susiana*, an Oriental species, is one of the strangest of hardy flowers—so weird, indeed, as to startle one on first beholding it. It is styled "mourning iris," its gray ground singularly and beautifully reticulated with dark purple. It looks like an Oriental flower; you find it some morning perched upon its stem, a great orchid on an iris stalk. Though it will withstand our severe winters with protection, and often without, its flowering is usually checked. It should be treated like the *tazetta Narcissus*, or stored during the entire winter. The iris, and, for that matter, all desirable and easily grown flowers, should be raised on a sufficient scale to afford an abundant supply for indoor use.

The *Pæonias*, including the tree, herbaceous, and Chinese sections, give us one of our most lavish floral displays. If you can not grow rhododendrons, these are excellent substitutes in limestone soil; they are equally floriferous, equally large-flowered, and equally varied in coloring. Earliest are the single dark crimson and the double fennel-leaved *P. tenuifolia*. The petals of the latter are a vivid scarlet-crimson, one of the most distinct reds of the year, its feathery foliage unlike that of any of its tribe. Roses are scarcely finer than some of the fragrant Chinese varieties, notably the pure white *festiva*, marked with carmine in the center, the dark-crimson Louis Van Houtte, the clear rose *Humeii* and Monsieur Boucharlat, and many others. Nor should we forget the old-fashioned red "piney," crimsoning in farmers' door-yards at the pretty things the great blue-bearded *fleur-de-lis* is telling her. The *Pæonia* may be said to grow itself, and, unlike the rhododendron, is perfectly hardy. Beautiful as a single specimen, massed in rows or beds few plants can vie with it for brilliancy.

I always rejoice when the azalea blooms. In it I find a charm presented by no other flower. Its soft tints of buff, sulphur, and primrose, its dazzling shades of apricot, salmon, orange, and vermilion, are always a fresh revelation of color. They have no parallel among flowers, and exist only in opals, sunset skies, and the flush of autumn woods. I admit that the rhododendron is magnificent where it can be acclimated; but, even in England and on the Continent, it is exceeded in gorgeousness by the azalea. Then, its delicious, uncloying perfume—why does not Piesse embody it in an essence? Its common name, —swamp pink—brings up its odor and its flame. A bed of azaleas with a foil of dark green is a sight worth going miles to see, and an acquisition worth obtaining at any price of peat and culture. The Ghent nursery-men who have developed its hues should receive a medal of rubies, topazes, and zircons, executed by a Cellini.

To the crossing of our common American species, *nudiflora*, *calendulacea*, and *viscosa*, with *A. Pontica* of southern Europe, and then selecting the best varieties raised from the seed of these crosses, we owe the so-called Ghent azalea. *A. mollis*, the Japanese and Chinese form, has been equally improved through hybridization and selection; these are smaller plants, with larger flowers. The azalea will not thrive in limestone soil, but should be grown in peat, or leaf-mold mixed with garden-soil, the soil well firmed about the plants. In the latitude of the lower lake region they require winter protection. With the azalea should be associated the native tall-growing lilies, *Canadense, Canadense rubrum,* and *superbum.*

A desirable border-plant is the columbine, or *Aquilegia,* in its many forms. Few perennials grow as easily from seed. They so very readily take crosses, however, that, where many are grown together, they can not be reproduced in the same character from their own seed. *A. chrysantha,* a Rocky Mountain species, with long-spurred, canary-colored flowers, and *A. cærulea,* with deep-blue sepals and white petals, from the same region, are the finest of the larger North American columbines. *A. longissima* is a species of western Texas, described as "flowers opening upward, spreading widely; of a pale yellow, or sometimes nearly white, or tinged with red." Its remarkable characteristic is its immense spurs, four inches and upward in length. It has been raised from seed in the Cambridge Botanic Garden, but has proved tender in that latitude. A contributor of "Garden and Forest," where it was recently figured, makes this interesting comment on its wonderful spur-formation : " In view of the recognized adaptation of flowers and insects to each other for mutual benefit, it is a question what long-tongued moths have developed side by side with this long-spurred flower, and how far the plant is really dependent upon such insects for fertilization." With the common scarlet columbine (*A. Canadensis*) almost every one is familiar. under cultivation it nearly doubles in size. There are numerous other American species, but none so fine as the Rocky Mountain forms. Many fine hybrids have been raised from these. A cross with the white form of *A. vulgaris* on *cærulea* has produced a flower of similar form to the latter, but of a pure snow-white color, two of these seedlings yielding double white flowers of the size and form of *cærulea*. These white forms, including the common white, are among the most beautiful of all. *A. glandulosa*, the Altaian columbine, and the scarce *A. Stuarti*, a hybrid between *A. glandulosa* and *A. Witmanni*, are pronounced the finest of the genus where they can be successfully grown, both requiring moisture at the roots, with perfect drainage.

Of the several kinds of *Hemerocallis* seen in gardens, none equals *H. flava*, the old-fashioned and always beautiful "yellow lily." Why the rusty-colored *fulva* should be cultivated at all, when there are so many better things to take its place, is beyond comprehension; yet country yards and city gardens are overrun with this coarse, spreading plant, whose flower is neither red nor orange, nor a good combination of both. It would require a gross of grub-hoes to eradicate it from the highway leading from any one village to another. Altogether a different plant is *H. flava*, frequently seen in country gardens. Indeed, the country garden often shows us the finest specimens; and I have sometimes thought, the more dilapidated the homestead and the larger the blue myrtle patch, the finer the golden clumps of the day-lily.

My garden was already generously stocked with this favorite plant, when, driving in the country, I saw two such uncommonly fine clumps growing in the unmown grass of a farm-yard, that the remembrance of them haunted me for days. I had no peace of mind until I should secure them. How they would light the front border! What vasefuls of cut blooms they would supply, without so much as being missed! An exchange for a dozen rose-bushes was the inducement I held out to the old lady who owned the coveted plants. The offer was accepted — not, however, without much persuasion; and the huge clumps, which one man could scarcely lift, were duly transferred to a post of honor. They threw up three spikes of bloom the following season! Perhaps they missed the chanticleer of the farm-yard to waken them into bloom ; perhaps they mourned the old lady's absence who had planted them and watched them and smelled them and complimented them, and given slips of them to her old lady neighbors—who knows? I may add that, since being transplanted, the plants have become re-established, and now flower with their former luxuriance. In these same tumble-down farmsteads flourish many a colony of the double poet's narcissus, which neither you nor I can grow under trees or in the open border half so successfully.

H. Kwanzo variegata is a large-leaved plant, attractive for its variegated foliage. *H. Kwanzo fl. pl.* is a robust species, preferable to *H. fulva*. *H. graminea* is a smaller *flava* in flower and foliage, and would be desirable were it not for its bad habit of spreading much at the root. I have found this almost ineradicable where it has obtained a strong foothold. The least particle of its white rootlets, under favorable circumstances, forms a plant if left in the ground, and it soon spreads and undermines its neighbors. None of the species equals the old-fashioned *flava*, one of the most satisfactory and beautiful of hardy flowers. It should be planted along the borders of a long drive-way, to realize its superb grace and beauty.

Another fine, old-fashioned, tall-growing perennial occasionally seen in country gardens is the fraxinella (*Dictamnus fraxinella*), so named from its pinnate leaves, resembling those of the ash. Its two forms, the pink-purple and the white, bear showy terminal racemes of larkspur-like flowers in June. Apart from its flowers and graceful foliage, its most attractive characteristic is the spicy fragrance of both leaves and blossoms. It suggests anise, sweet-clover, and lavender.

So powerful is the volatile oil generated by its flowers, that a lighted match held several inches above the plant, on a still, hot summer's evening, will cause a flame to appear. A native of the Levant and southern Europe, it may be increased both from seed and root-division, the former being preferable. You should plant it along your favorite walk, with the lemon-balm and the anise-scented giant hyssop, so that you may pluck a leaf of them as you pass.

I see, in many an old homestead along the shaded highway, the prim box-hedge inclosing the garden of old-fashioned flowers. Often as the swallow returns do they rise anew and blossom with perennial freshness. The flowering locust-trees, and the tansy-bed running wild outside the fence, give a hint of the fragrance within, where I see the water-bucket ready for its floral libation. I push open the wooden gate, to be greeted by the first snow-drops, the daffodils, the yellow crown-imperials, the grape-hyacinths. I see the blue irises, the larkspurs, the bell-flowers, the bachelor-buttons, the monk's-hood. I note the big double white poppies, the clumps of sweet-clover, the drifts of snow-pinks, the white phloxes. I see the *Dielytras*, the sweet-williams, the tall, yellow tulips, the sword-grass and ribbon-grass, and *Tradescantia*. I smell the sweet-peas, the valerian, the madonna-lilies, the white and purple stocks. I inhale the breath of the lilies of the valley, the brier-rose, the white day-lily, and the purple wistaria twining about the porch. I see, too, the double-flowering rockets, the spotted tiger-lilies, the dahlias, the rows of hollyhocks, and the phalanx of sunflowers.

Then, the flowering shrubs of the old-fashioned garden—the snowberries, honeysuckles, and roses of Sharon, the storm of the snowballs, the mock-oranges, and the great white lilacs leaning over the hedge, heavy with their blossom and perfume. Nor is the herb-garden of the Fourth Georgic forgotten, where

> Cassia green and thyme shed sweetness round,
> Savory and strongly scented mint abound,
> Herbs that the ambient air with fragrance fill.

Here grow mint, marjoram, anise, sweet-basil, catnip, lavender, thyme, coriander, summer-savory, and, last but not least of the fragrant labiates, the pungent sage, that will ruin the dressing of many a Thanksgiving turkey. A sassafras-tree not unfrequently grows, by accident or design, somewhere about the yard; and there is sure to be a red horse-chestnut, or a trumpet-flower, for the humming-birds to plunge in.

How the swallows wheel and dive over the weather-beaten barn, and twitter among the eaves they have visited generation after generation! And what a honey-laden wave surges over the neighboring clover-field! I recall such a farmstead on the crest of the Livingston hills, where farm-life always appears at its pleasantest. All around it extends the panorama of wood, ravine, and purple upland, changing with every change of atmosphere, open to every effect of sun and cumulus-cloud. Here, I thought, a philosopher might find the coveted stone. Life always seems so restful and its current so placid on the summer hills. But we forget the blighting frost, the moaning blast, the wintry shroud. In life, things are pretty evenly balanced, after all; and while summer is delightful in the country, to the most of us, in winter, it is pleasanter to think of in the city. Those who really love the country in its harsher aspects are few. I doubt if there exists another Thoreau for whom "the morning wind forever blows, bearing the broken strains, or celestial parts only, of terrestrial music."

I see, too, the neglected farm-garden; one passes many such along the dusty road. Here, an old locust and mock-orange have been allowed to sprout at will; the blue iris has crept outside the fence, with clumps of double daffodils turned over by the plow and flung on to the road-side. There, is a jungle of stunted quinces and blighted pear-trees. The spreading myrtle-patch has usurped the place of what was once a lawn; tall thistles, hog-weed, pig-weed, and burdocks make and scatter seed year after year; an army of weeds has overrun the path— the plantain, purslane, goose-grass, dandelion, joint-weed, and mallow; and a green goose-pond, over which are hovering yellow butterflies, exhales its miasma in the sun. Once the garden was beautiful, famous for its old-fashioned flowers, and many are the "slips" the neighbors obtained from its floral stores. The grain-fields and fat pastures corresponded with the luxuriance within. But the farm changed hands on the death of the owner, and the new owner cared little for the flowers, and has left the farmlands mostly to themselves. I always hurry by the farmstead; its dilapidated out-buildings look as if they might be haunted by the ghosts of starved and neglected animals.

As I stroll through the garden toward evening, I find the brown May-fly has suddenly appeared in legions. Every bush and tree swarms with them; while, high as one can see, the air is throbbing with their undulating flight. Now up, now down they go, flitting on wings of gossamer, their antennæ and long tails balancing them in their graceful dance of an hour. Is it simply to gorge the bats and the trout, which make the most of the insect-manna, that the May-fly is sent?—for the naturalists do not ascribe a cause for its brief existence, in the reason of nature.

The first of the innumerable young broods of sparrows are fledged, and have begun their interminable shrieking. The foliage is so thick that it is almost impossible to shoot them; and to attempt poisoning them is out of the question, on account of the few remaining song-birds. How wretched they render human life! What a constant burden for the ear to bear! If they would only mew like the cat-bird, or do anything to vary the tedium of their incessant "Cheep!

cheep!! cheep!!!" I envy the deaf, and the fat men who drown all other sounds with the sound of their own wheezing. My neighbor's parrot, who yells like all the fiends of Dante's Inferno, has at least the merit of variety in his voice. If the sparrow continues to multiply, there will be a new verdict rendered at coroners' juries; his monotonous cry is fast abbreviating the allotted span of mankind.

Meanwhile, the floral procession is advancing in the flower-borders. The large Oriental poppies are rightly named, and, with their fine foliage and immense flame-colored blossoms, are undoubtedly the most gorgeous of garden-flowers. You could almost light your pipe from them. The variety *bracteatum* is the stouter grower, holding its stalks more firmly and erect, and is the superior in the color and beauty of its lustrous, dark scarlet flower. The petals of the Oriental poppy are oddly marked with purple-black spots inside, forming a black cross. Parkman's Oriental poppy, originated near Boston, is another fine form, as yet rarely seen. The Oriental poppies and the yellow day-lily, blossoming at the same period, should be largely employed in the border and other suitable places of the garden.

I have planted the tall, late-flowering tulips

freely among the poppies, the luxuriant foliage of the latter concealing the naked base of the tulips. A mass of tulips thus grown produces a much finer effect than when bedded by themselves. The tulip invariably looks better in neglected gardens for this reason; it is seldom seen rising from the bare earth, generally springing from the grass or shrubbery, or at least having a background of green. Seeds of these big Orientals should be sown in February in the greenhouse, so that they may germinate early, be pricked off, and form strong plants to set out as soon as possible in May. While they are perfectly hardy, small plants are generally winter-killed. I find growing among my *P. bracteatum*, raised from seed, a distinct variety with smaller flowers of a peculiar and very beautiful cherry-red.

We must go to the Orientals to learn the true use and significance of flowers. "Very beautiful are the flower-customs here," says a writer from the lands of Kâlidâsa and Fírdusi. "In Bombay, I found the Parsees use the Victoria Gardens chiefly to walk in, 'to eat the air.' Their enjoyment of it was heartily animal. The Hindoo would stroll through them, attracted from flower to flower not by its form or color but its scent. He would pass from plant to plant, snatching at the flowers and crushing them between his fingers as if he were taking snuff. Presently a Persian, in flowing robe of blue, and on his head his sheep-skin hat, would saunter in, and stand and meditate over every flower he saw, and always as if half in vision; and when the vision was fulfilled, and the ideal flower he was seeking found, he would spread his mat and sit before it, and fold up his mat again and go home. And the next night, and night after night, until that particular flower faded away, he would return to it, and bring his friends in ever-increasing troops to it, and sit and play the guitar and lute before it, and they would all together pray there, and after prayer still sit before it, sipping sherbet and talking late into the moonlight; and so again and again every evening, until the flower died. Sometimes, by way of a grand *finale*, the whole company would suddenly rise before the flower and serenade it with an ode from Hafiz, and depart."

I suppose we could not do without the June *Pyrethrum*, it is so floriferous, and has such feathery, deep-green foliage. Nevertheless, I see no excuse for littering up a garden with some of its crimson-magentas or magenta-crimsons. Weeded of its bad colors and bad centers, it is certainly worthy of all praise. It lasts long, and its flowers are excellent for cutting.

Speaking of bad colors, I think there is much in what a young lady once observed to me at a ball, the conversation turning on the newly decorated rooms. " I don't think the glaring combinations and unhappy uses of color we frequently see in houses and exhibited in dress so much the fault of individual taste as of a deficiency of the color-sense. Let us count the green dresses, of which there seem to be an unusually large number present, and I assure you in advance that at least every third person you ask will pronounce the delicate shades of green blue. It is the same with reds. A hideous solferino looks all right to some; it appears the same shade to them, doubtless, as a cardinal or a terra-cotta or some other shade does to you. I haven't the slightest doubt that color-blindness is at the bottom of much of the distress that one's eyes are forced to encounter." Solferino and magenta, or shades closely touching upon them, should not be tolerated in the garden. They are weeds, that ought to be eradicated as soon as they appear.

A writer in the London "Garden" gives a simple rule to determine whether colors harmonize: "People who have no natural perception of color can not be trained to arrange colors harmoniously by any code of rules; but those who have a natural feeling for color can find out whether any two colors harmonize by a very easy test. Place the colors separately on a gray, white, or black ground. If they are brighter, richer, and fuller together than separately, they harmonize; but if not, they should not be placed together."

I could say more in favor of spiræa or *Hoteia Japonica* were it not so susceptible to the hot sun. Charming so long as it remains fresh, during average seasons its foliage is soon blighted and its beauty destroyed. The hardy, large perennial spiræas are beautiful with their graceful spikes and plumes and panicles. Of these, *S. aruncus*, the familiar goat's-beard, is among the finest when well established and allowed sufficient room to attain its full development. *S. Humboldtii* is equally robust, though its flowers are not so pure a white. The species *filipendula* and its double are worthy a place in the border if only for their graceful, fern-like foliage. *S. ulmaria fl. pl.*, and its form with golden variegated foliage, are both desirable species. The prairie *Spiræa lobata*, with its rosy carmine cymes, must take the place of the finest of all the meadow-sweets, the Japanese *S.*

palmata, which does not thrive in this climate. Under cultivation, all the herbaceous spiræas prefer partial shade, and, to appear at their best, should be supplied with water in hot weather, or their appearance is soon marred by withered foliage.

What would the old-fashioned garden do without the sweet-william; and the new-fashioned one, too, for that matter? It is as indispensable as the snow-pink, the carnation, and the aster. "*Die fallen ins Aug'!*" they fall into the eye, to quote from the gardener once more, an apothegm I think worth embalming. Gay they are, with their infinite colorings and their prodigality of bloom. The *Dianthus* are all of them pretty, notwithstanding the interminable appellation of one, *Dianthus sinensis Heddewigi diadematus flore pleno!* Leave them alone, and they will sow themselves; sow the seed on good ground, and they reward you a thousand-fold. They vie with the auriculas in their merry eyes, and are almost as brilliant and fourfold as lasting as the poppies, unless I except the Icelanders. Even the old maids love their sweet-williams. In Gérarde's day it was "esteemed for its beauty to deck up the bosoms of the beautiful, and garlands and crowns for pleasure." It is well to caution those who grow it, however, not to place it close to gravel-walks, where the seeds are apt to drop and cause no little trouble; they germinate so freely. Seeds should be saved from the best kinds, most desirable colors, and strongest trusses. The *Eschscholtzia* and *Coreopsis* become almost a pest unless the seed-capsules are cut off, and the *Calendula* is also troublesome in this respect; but the latter may be excused, it is so warm and steadfast in late autumn when we could hardly do without it for cut flowers.

With the sweet-william is often associated another old garden favorite, the snow-pink (*Dianthus plumarius*), a charming subject when well grown. I once saw an immense patch of this in front of a country cottage, growing so luxuriantly that the fragrance drifted far out on to the highway. I stopped to inquire of the *genius loci*, who was busy with her watering-can, how she grew them so finely and so profusely. "I pinch them, give them plenty of water, and keep up a fresh stock from cuttings every two years." The old story, I thought; new words to the old tune—"care."

The large bell-flowers are doing excellent duty as flowering-plants, notably the old-fashioned Canterbury bells (*Campanula medium*). Their immense scalloped goblets of diversified colors are preferable in the single to the double and duplex forms. The several spikes are stout and the species is of robust habit, altogether a grand border-plant. So also is the strong and taller *C. macrantha* with blue-purple bells. It should have a partially shaded place in the back row of every border. The peach-leaved bell-flower (*C. persiscifolia*) is an excellent border-plant, but does not hold itself erect like the other species, and therefore needs staking. All plants, it may be observed, that require support should be staked early, instead of being left until they begin to flag. For supports iron stakes are the neatest. The Austrian harebell (*C. pulla*), a small species with lovely, drooping purple bells, would be an admirable subject for the rock-garden were it not for its rambling root-stalks. *C. barbata*, the bearded harebell of Switzerland, I have not found as satisfactory as some of its relatives.

To judge from the description and illustration, the finest of the bell-flowers—if it may justly be termed a bell-flower—must be the Bokhara bell-flower (*Ostrowskia magnifica*), just introduced into Europe, a grand chime of bells crowning a tall, leafy spire. The stem is stout, from three to five feet high, the leaves in whorls; and the flowers, which are five inches or more in diameter, placed in loose terminal panicles—pale mauve varying to light blue, with a large, club-like stigma. A country that can produce such rugs as Bokhara, acquiring with time a color and bloom like that of a ripe peach and plum, ought to contribute an extraordinary flower; but whether the flower will improve with age and wear in a foreign climate is as yet undetermined.

I do not hear anything of the great Californian poppywort (*Romneya Coulteri*), which created such a stir on its introduction into England. Mr. F. A. Miller, of San Francisco, who introduced it twelve or fourteen years ago, wrote me, "There is no flower that combines so many good qualities—such a fragrance, beauty, and general effect—as this plant." Unfortunately, it will not survive our rigorous climate, and I believe it has failed to establish itself in most gardens where it has been tried in England. In her plants California is not accommodating, as a general rule, Nature having for the most part suited them only to the climate of their birth. They are ill adapted to our sudden snaps of winter returned.

The roses are now in their prime. I had occasion to cut a collection this morning—June 22d—rising shortly after three o'clock. A rustling of the tree-tops was the first precursor of dawn—the breeze which nearly always precedes awakening day. At 3.20, before it was yet light, the cat-bird was first of the songsters to salute the morn. Five minutes afterward the wood-pewee drowsily voiced the first two notes of his refrain —"whē-ū whē, whēē-ū!" In just two minutes more a robin began his matin song, followed by the crowing of the cocks, which quickly ceased, until at 3.40 the wood-pewee began whistling merrily, immediately succeeded by the robins, wood-thrushes, sparrows, and various song-birds, all joining in the morning chorus. At four the *crescendo* was at its height, when it gradually diminished, soon leaving the sparrows in almost undisputed possession. I found the honey-bees busy among the raspberry-blossoms a few minutes after four, and the big bumble-bees but a little later to begin their morning task. Of all these early risers I for once was the earliest.

The hollyhock may be termed a great power in July. Classed as a biennial, it might almost come under the head of perennials, being as permanent as many true perennials. It was a favorite of Wordsworth and is also of Tennyson. Tennyson's summer,

. . . buried deep in hollyhocks,

is expressive of the luxuriance of this Chinese flower. It should be seen in long rows, in well-drilled color-columns, to exhibit its most striking effect, each plant a sentinel in uniform, and each with rosettes brighter than his fellows. The hollyhock will grow anywhere; it will grow doubly well with deep cultivation, and when liberally manured and watered during dry weather. Dampness being injurious during winter, it is recommended to remove the earth about the crowns in autumn and fill up with six inches of white sand. Propagation is effected from eyes, seeds, cuttings, and division. The thrip and red spider are fond of the hollyhock, and hence the rusty appearance so many plants present. If you have four or five gardeners, this may be obviated by syringing every leaf, upper and under side, of the long rows daily with whale-oil soap and tobacco-water. The hollyhock also demands an admission fee.

The graceful spring bitter-vetch (*Orobus vernus*) is past its flowering, but still retains its handsome foliage. *Hieraceum aurantiacum* has passed, after showing its peculiar orange-red flowers, even more odd in color than those of the native orange-red milkweed that stains the sandy places in midsummer. The creamy trusses of the tall valerian are a hive of sweetness, and the yellow camomile (*Anthemis tinctoria*) is covered with its daisy-like flowers, rejoicing in the increasing heat. It will soon be succeeded by *Coreopsis lanceolata*, another of the showy yellow composites, with the ever-blooming pea, the double-flowering rocket, and the large-leaved day-lilies, of which *Sieboldii* has the finest foliage, and the white variety the finest and sweetest flower.

Not without just reason is the larkspur included among the nine flowers specified in the garden of " Maud "—the woodbine, jasmine, violet, acacia, pimpernel, rose, lily, passion-flower, and larkspur. Keats should have included it in his sonnet on blue. Holmes alludes to it neatly in the " Autocrat ":

Light as a loop of larkspurs—

light in its poise, and light or dark, as you wish it, in its complexion, and beautiful in all its forms.

Sauntering at dusk through the fragrant garden alleys, I hear as in a dream the last faint notes of the vesper-sparrow; and see, kindling the edge of the covert and sparkling amid the shrubbery-glooms, the myriad fire-fly revelers merrily dancing out the last sweet night of June.

Two Garden Favorites.

I love the lily as the first of flowers,
Whose stately stalk so straight up is and stay.
                    ALEXANDER MONTGOMERY.

            . . . The coming rose,
The very fairest flower, they say, that blows,
Such scent she hath ; her leaves are red, they say,
And fold her round in some divine, sweet way.
                    PHILIP BOURKE MARSTON.

VII.

TWO GARDEN FAVORITES.

ALPHABETICALLY, the lily comes before the rose; and in the summer-garden, which would lack its greatest charm if deprived of either, the common orange-lily appears before the first June rose.

Is this significant; and shall I say the flower singled out in the sixth chapter of St. Matthew's Gospel excels its sister in floral graces and virtues? The rose, as we generally admire it, as it is eulogized by the poets, is a florist's flower. Its rival, equally well known, and almost if not as freely extolled in poetry, owes less to man and more to nature. I would not detract from the rose, when I say it is less graceful than the lily and its form more artificial. In comparative merits of color and fragrance it would be difficult to discriminate; each has its claims that may not be overlooked. I may add, on the other hand, if you smell of a lily you are liable to be stained by its pollen; and if you pluck a rose, there lurks the hidden thorn. Perhaps the lily and the rose, or the rose and the lily, furnish a case in point where comparisons are odious, and each one may better decide for himself which is the superior flower.

I begin with the lily, therefore, because it comes first alphabetically, and is first to appear. Whispered the white lily to me: I am the emblem of purity, the type of saintliness; at the altar and at the tomb I bring joy and consolation; in the garden I am sweet beyond all my companions, and with my whiteness none can compare; I am sweet, I am chaste, I am beloved by all. Do you know my origin? "Jupiter wished to make his boy Hercules (born of a mortal) one of the gods: so he snatched him from the bosom of his earthly mother, Alcmena, and bore him to the breast of the god-like Juno. The milk is spilled from the full-mouthed boy as he traverses the sky (making the Milky Way), and what drops below stars and clouds and touches earth, stains the ground with lilies."

So extensive and beautiful is the genus *Lilium*, so varied in form, color, and periods of blossoming, that, like the daffodil, a garden might be composed of it alone. We readily concede its beauty; the next thing is to manage it. "The more I see of lilies, the less I know how to grow them," is a wise maxim of H. J. Elwes. One requires tact and perseverance to grow the lily. Very many of its numerous species are fastidious, quick to express their likes and dislikes; some, indeed, refuse to yield to culture unless in a climate of their own choosing. Yet, after all, most of the species may be satisfactorily grown if proper attention be paid to soil, position, and protection.

While the majority of the genus are hardy, and very many are natives of cold climates or high elevations, winter protection to nearly all species is nevertheless advisable with us. If the ground remained covered with snow the entire winter, the bulbs would not suffer. It is the alternate and frequent changes from freezing to thawing which contract and heave the ground that causes the trouble, the bulbs themselves contracting and expanding with the changes of temperature. No less important is the matter of drainage: very few lilies will endure being water-logged; very few, also, will endure manure about their bulbs. The manure harbors wire-worms, which are fond of the lily's tender scales. To obviate this, and to strengthen root-action, all lilies, on being planted, should receive a liberal sprinkling of sharp sand about the bulbs.

With us the lily is even more susceptible to drought than to frost, and failure is oftener the result of shallow planting and poor soil, than owing to the rigors of our winter climate. Very much depends on good, deep, and congenial soil, and healthy bulbs to start with. Partial shade, with some species, is absolutely necessary, and all are benefited by, and some will not grow at all without, a liberal supply of moisture. Different species are as different in their requirements as they vary in the character of their bulbs and their periods of flowering. What holds good of one climate often does not of another. I have seen magnificent beds of established *Lilium auratum* and *speciosum* on the Eastern coast in open sun, that it would be utterly impossible to grow without shade in the lower lake region. They liked not only the peat and deep trenching, but extracted a tonic from the sea-air, which just met their requirements. It is one thing to grow certain plants where the climate itself grows them; it is quite another thing where they have to be cajoled into tractability. The more difficult the task, however, the greater the satisfaction to accomplish it; success is always pleasant, whether to grow a capricious flower or banish a troublesome weed.

The following is the last classification adopted by Mr. J. G. Baker in his "Synopsis of all the Known Lilies," published in 1875:

I. Subgenus CARDIOCRINUM (leaves heart-shaped). Types: *L. cordifolium, L. giganteum.*

II. Subgenus EULIRION (flowers funnel-shaped). Types: *L. longiflorum, L. candidum, L. Washingtonianum.*

III. Subgenus ARCHELIRION (flowers open). Types: *L. tigrinum, L. speciosum, L. auratum.*

IV. Subgenus ISOLIRION (flowers erect). Types: *L. croceum, L. concolor, L. Philadelphicum.*

V. Subgenus MARTAGON (flowers turban-shaped). Types: *L. martagon, L. superbum, L. pomponium, L. polyphyllum.*

If one would go distracted on the subject of forms and varieties, he should peruse the annotated "Alphabetical List of the Species and Varieties of Lilium," compiled by M. d'Hoop, a Belgian amateur, published in vol. xxvii, No. 692, of the London "Garden." Thus, under *L. Canadense*, its principal varieties are described as *L. C. superbum* (intermediate between *Canadense* and *superbum*), *L. C. rubrum*, *L. C. Hartwegi*, *L. C. minus*, *L. C. occidentale*, *L. C. parviflorum*, *L. C. parvum*, *L. C. puberulum*, *L. C. Walkeri*. No less than seven forms of *L. Philadelphicum* are mentioned: *L. P. andinum*, *L. P. wanscharicum*, *L. P.* of Brentwood, *L. P.* of Connecticut, *L. P.* of Massachusetts, *L. P.* of the Orange Mountains, *L. P. varietas Hookeri*.

First among the lilies is one of the three most common and easily-grown species, the tall orange-lily (*L. croceum*). This would show to better advantage if it did not appear with the Oriental poppies, which overpower everything else in red about them. The orange-lily looks best springing from the shrubbery, and, like the tiger-lily, needs to be seen in strong, well-established clumps, to show its real characteristics. The orange-lily is succeeded, a few days later, by one of the finest of lilies, the Caucasian *L. colchicum*, much less frequently seen than its merit deserves—a soft canary-yellow flower, speckled with small dark-brown spots on either rim of the petals, and exhaling an intense and individual odor. It is a slow species to arrive at perfection, and, owing to the cernuous habit of its flower, is not seen at its best until well established and its stems rise to their full height. As it blossoms with the conspicuous lemon-yellow day-lily, it should be placed where it may be seen by itself. This species varies not a little in the character of its flowers, some being larger and deeper-colored than others, and having the petals more freely spotted; it is one of the easiest of lilies to raise from seed. *L. colchicum* does well in the open sun, but grows larger in partial shade, where it also holds its flowers and foliage better.

The voracious rose-beetle is becoming more and more omnivorous. Prompt to appear with the first white Madame Plantier rose, his armies soon pounce upon the white pæonias, which would be utterly ruined were he not kept in check. Last year he added the white Iceland poppy and *Spiræa filipendula* to his bill of fare, and to-day I find him attacking the colchicum lilies. One can not gather a bucketful and toss them into one's neighbor's garden, for they would only fly back again. My neighbor, who lets his chickweed and dandelions go to seed, is, I think, the main cause of their increasing numbers, for he never lifts a finger to destroy them.

Siberia contributes one of the smallest and earliest of the lily family in *L. tenuifolium*, prettier as a cut-flower than when growing out of doors, where its many wide-branched blooms and sparse leafage on slender stalks give it a top-heavy appearance. Its small vermilion, wax-like, and strongly-scented flowers are distinct among the turbans.

L. pulchellum, another small red species, from Siberia, blossoms with *tenuifolium*. Both of these do best in sandy soil, as does also the common wild orange-red lily (*L. Philadelphicum*), a most beautiful early species. You have seen its single and sometimes two and three flowered blossoms lighting the June meadows and sandy hill-sides. Its blossoms seldom number more than three. A gigantic specimen I once found with eight blossoms, and which I carefully transplanted with a large ball, divided itself into four stalks the following season.

I do not wonder that the Madonna lily (*L. candidum*) has been claimed as an emblem by nearly a hundred saints. It seems to have a special charm of its own, so chaste it is, so inviolable in its purity. The roses and the big blue larkspurs come into bloom just in time to set it off, and together, perhaps, form the most beautiful summer pageant of the garden. The Madonna lily is one of the most gracious of its graceful tribe, being not only unusually hardy, but quick to increase, and thriving in almost any soil and position. Though its white print is seen everywhere, it is a flower that is never common. One of the easiest to grow, it is no exception to the rest of the genus in its dislike to being disturbed. The right way is to think twice before placing any plant or tree, so that, when once planted, it will not be necessary to interfere with it. Where transplanting is necessary, the lily should be moved when its bulb is at rest—a period easily determined by the dying down of the foliage and stalk. Many lilies require several years to become established, and, so long as they remain healthy and flower well, they should not be disturbed. What applies to the daffodil does not hold good with the lily; and I think the rule laid down by many, that the latter is benefited by transplanting and dividing every two or three years, is wrong. None of the varieties of the white lily can compare with the type; the double form is as great a failure as the rose-colored lily of the valley.

The past year the white lilies were not as fine as usual, something in the late spring, or else the previous dry autumn, affecting them. The stalks were less strong, and the leaves often turned yellow before the appearance of the flowers.

The lily should not have its stalk cut down after blossoming, until the leaves have fallen off, and the stalk becomes yellow and shriveled. It is always a temptation to cut down the withered stems, which are unsightly. But to remove the green stems means to make the bulb go to rest prematurely, the result being that the next season the flower-stems come up weaker and produce smaller flowers. There is no objection to cutting the stems down gradually from the top as they become dry; this does not weaken the bulbs, and at the same time avoids the appearance of untidiness.

We would naturally expect much of the scarlet martagon or scarlet Turk's-cap (*L. chalcedonicum*), the true "lily of the field." Indeed, it is never disappointing, except when it is disturbed, the species being extremely sensitive to removal, and never being good for several years after transplanting. It is one of the grand things in red; an old clump of it, in fiery scarlet flower, is a sight for a cardinal to dream of and a humming-bird to admire. Its cultural requirements are as simple as those of the Madonna lily, and the beautiful cross between these two, the Nankeen lily (*L. excelsum, L. testaceum, L. Isabellinum*). No garden should be without this fine hybrid to accompany the white lily. It inherits the stateliness and the combined perfume of both parents, with a soft apricot or buff-salmon color unique among its family.

An overestimated lily, I think, is the yellow *L. Hansoni.* It is to the Japanese species what the panther lily (*L. pardelinum*) is to the North American kinds—there are many finer to choose from. But both are easy to grow, and the grand whorls of *Hansoni* certainly are not to be despised. Its small turban is of a distinct yellow, with a peculiar Oriental odor—you would know it came from Japan with your eyes shut. I should, doubtless, admire it more if I could grow it larger. I place it above *pardelinum*, which passes by quickly, and has a loose sort of flower on limp stalks that always require support; Montgomery would never have grown the latter in his lily garden. The Californian *L. Washingtonianum* is, I think, also overestimated—difficult to grow, and very fleeting. *L. Humboldtii*, *L. rubescens*, and *L. Parryi* are finer. All the Californian species, except *pardelinum*, are more or less difficult to manage; they often remain in the ground a long time before appearing. These do better in some portions of England, where they are consigned by the thousands, to be sold at auction. No little confusion has existed concerning the Californian species. There are differences in plants which florists readily recognize, but botanists will not. Thus *L. rubescens*, one of the handsomest of the species, was formerly classed with *L. Washingtonianum*, a distinct species in almost every particular.

Recently a yellow form of *pardelinum* has been discovered, together with another species, which the discoverer, Mr. F. A. Miller, of San Francisco, informs me he has designated as *L. pardelinum Alpinum*. This, he states, "grows on dry ground, and in general characteristics is not unlike *L. parvum*, which, however, only grows on very wet ground, or along water-courses. The flower is small, but vivid and rich in color; nearly half of the flower, which appears horizontally, is scarlet. I found it at an elevation of eleven thousand feet, higher than the altitude where any lilies grow usually."

Where it can be well grown, *L. speciosum*, with its numerous varieties, is unquestionably one of the finest of the genus. The Massachusetts climate, which produced the beautiful variety Melpomene, suits it; but it is usually seen at its best under glass. *L. Brownii*, another Japanese species, is far more rare, but scarcely as handsome as the common *L. longiflorum* and its varieties. Contrary to general opinion, I have found the former extremely slow to recover after lifting. *L. Harrisii*, the Bermuda lily, is best suited to the greenhouse, on account of its tendency to start so early, and is not to be compared with the Japanese long-trumpeter for out-of-door culture. An easily-grown lily is the European

Turk's-cap (*L. martagon*), and its fine varieties, *album* and *dalmaticum*; the latter is said to revert to the type after a few years' cultivation. There are scores of varieties to choose from in the Japanese species *Thunbergianum* or *elegans*, nearly all of which are dwarf in habit, and vary in color from pale apricot, orange, and orange-red, to blood and deep red. These are among the easiest of the genus to grow, and do not like shade. *L. bulbiferum*, somewhat like *Thunbergianum*, with orange-crimson flowers, is also one of the least fastidious species; the variety *umbellatum* is a stronger grower than the type. Both of these are valuable early species where a mass of red in lilies is desired in open sun.

Of the many species we owe to Japan, none can compare with the great golden-banded lily (*L. auratum*) and its varieties; if, in reality, it is not the finest of its tribe. But it is a coquette at heart, and, unless wooed earnestly and persistently, in ninety-nine cases out of a hundred it will only smile bewitchingly the first year, to jilt you the next. Of the hundreds of thousands of bulbs imported annually from Japan by Europe and America, very few remain after the second and third year. This is not owing to its tenderness, for it is among the hardiest of the genus. Neither is it a mere question of climate and culture. Climate and culture have much to do with it, but the main reason of its failure is beyond this. Investigation has only recently brought to light the chief cause of its disappointment. In its own home it is infested by a mite, which, however, does not seem to cause trouble until it leaves its native country. The enfeeblement incident to the removal of the bulb, together with the difference of soil and climate, cause its deterioration. Some unusually strong round bulbs, which may not be so much affected, if placed amid congenial surroundings, are able to resist this tendency; and it is only by selecting a quantity of the best bulbs to start with, and retaining the most robust of these after the first year's flowering, that we may hope to establish this lily; that is, unless it can be grown more successfully from scales or seed, a process seldom tried in this country, where we have not the patience to wait. Of fifty bulbs, perhaps only one third, more frequently a quarter or less, remain after the second year, even when grown under the most advantageous circumstances. This is what the term "home-grown *Lilium auratum*" means, or is supposed to mean; for the loss is always so great that few care to deal in *auratum* bulbs, except as directly imported.

Notwithstanding this, so desirable is the golden-banded lily, that it is worth any amount of trouble to establish. Peat, with the addition of sharp sand, seems to meet its cultural requirements best, although it does well among Onoclea ferns, in soil largely composed of black "muck" or decayed wood. A sufficiency of water it must have, and abundance of shade is absolutely necessary to success. The midday sun is fatal to it. A flickering shade, I should say, is best for this and most lilies. It should also be placed where it will not be subject to high winds. The *auratum* is one of the most protracted of the genus in its flowering period, and scarcely two of a number of bulbs planted at the same time come into flower simultaneously. There are numerous varieties of this species, all of which are beautiful; the more pronounced the terracotta spots and vivid the color of the ray or central band, the finer the flower.

I regard a well-grown *Lilium auratum*, with a strong stalk rising to a height of five or six feet, supporting its dozen or more deliciously-scented blooms, as the grandest of all hardy flowers. It is worth planting a hundred bulbs to establish one such embodiment of floral beauty. When I stand in its lovely presence I am repaid for any trouble; and I freely forgive the Japanese all the misery they have inflicted upon Kiota and Awata. It is scarcely astonishing that a country which can produce such a flower should produce artisans to whom nothing is impossible. It ought to inspire a transcendental literature. Under date of August 29, 1885, F. Bridger, Penshurst Place, Kent, wrote to the London "Garden": "We have in the open ground here a *Lilium auratum* with forty flowers upon it at the present time, and over a hundred more still to open; the plant has six stems seven feet high." The proprietor should go down upon his knees to such a gardener, and endow him with an annuity for life!

Remarkable among lilies, and differing entirely from the type, is the Himalayan species, *L. giganteum*, termed the "king of lilies." It is, I believe, generally considered tender with us, and difficult to manage. Two years ago I experimented with three of a dozen small bulbs, planting them out on the 20th of November, in rich loam and leaf-mold. These wintered perfectly, and the remainder, which were placed in a cool house, have since withstood the winter equally well, and are now vigorous plants, with immense *Caladium*-like leaves, growing in partial shade; these have not yet flowered. This species, in Europe, attains a height of ten feet, and bears huge trumpet-shaped, nodding white flowers, interiorly stained with purple, and of powerful fragrance. It is a strong rooter, and, as it pushes up very early, it should be planted rather deeply, and protected with fine ashes from spring frosts. It is said to require years before it sends up its flower-stalk, and the longer it is in coming into flower the finer it is said to be.

The tiger-lily (*L. tigrinum*), an occupant of most gardens, is never common when well grown. Its odd Chinese color and pronounced spots must be seen in mass to do it justice; the old-fashioned country garden invariably does well by it, because it is left undisturbed. *L. tigrinum splendens* is termed the most beautiful, though the double variety is almost equally fine. All of the tigers are among the very easy lilies to grow.

A single specimen of a beautiful native lily of the *Canadense* section was discovered in 1840 by Dr. Asa Gray, on the Alleghanies, North Carolina, and named in his honor *L. Grayi*. This is described as having flowers of dark-red orange, uniformly dotted within with rather small purple spots. Although since found in the same habitat, the species is as yet extremely rare.

A lily distinctly American is the wild Turk's-cap (*L. superbum*), an inhabitant of meadows and low grounds, the tallest and most numerous flowered of our native lilies. So variable is this in its size, shape, color, markings, and the number of its flowers, that it is difficult to specify it distinctly. It is a question, moreover, just when it becomes concurrent with *L. Canadense rubrum*, as would not unfrequently seem to be the case. The most common forms of the species bear dingy red or yellowish-red flowers, and vary greatly in the robustness of the plants. *L. superbum*, as usually sent out, is anything but the superb lily it is in certain favored localities, and none who have only seen its more common forms have any conception of its stately beauty in its rarer and perfected state. Along the Old Colony Railway, between Newport and Boston, and on the Shore Line between New London and Boston, the species is seen at its best. For miles it follows one along the railway, steeping whole meadows in scarlet, the color of the flowers varying from the most intense bright crimson to dingy yellowish-red. There in the salt air it revels even on dry, poor soil, bearing from three to fifteen or more commonly three to seven flowers on a head.

In its cultivated state, where well grown, the large form is still more free flowering, the bulbs throwing up from a dozen to three dozen blooms on stems eight to nine feet high. I have never seen it as brilliant under cultivation as it occurs wild in the localities referred to. Neither have I ever seen the lemon-yellow *Canadense* as vividly colored or as tall as it occurs near New London, Conn. Fine color and tall stalks with *L. superbum* under cultivation, however, will largely depend upon good selection made in the native habitat. This year a disease seems to have affected *L. superbum* under cultivation in some places, causing the stems to shrivel and the leaves to rot off.

Of the graceful Turk's-caps or turbans there are none, I think, unless I except the rare form of *L. superbum*, equal to the red *Canadense*, our own wild wood-lily. I know of no lily more graceful or stately. It is as distinctly American as the cardinal-flower or the pink lady's-slipper. Something it possesses of the wildness, the suppleness, and the charm of cool leafy places—in its tall, polished wand, its fluttering whorls, and the pure whiteness of its rhizome. It always looks self-possessed, bending but never breaking before the rain and storm. Then its life and fire when rising from the foil of light-green Onocleas. I find it growing in low woods where water has lodged in spring, lifting its lithe stem along shaded ditches and hedges, and rising in flexile grace amid the chosen haunts of the sensitive fern. Owing to its increased vigor the red form of *Canadense* is preferable to the yellow, though the latter is exquisitely beautiful in the color and poise of its flower. Certainly the yellow form of *L. Canadense* far surpasses any yellow form of *L. superbum*, the latter invariably having a washy appearance.

L. Canadense rubrum is much earlier to blossom than *superbum*. The distinction of shape of flower, however—*superbum* being quite recurved in the Turk's-cap style—is, perhaps, more obvious than any other characteristic. I find the red *L. Canadense* extremely protean, plants of similar size occurring side by side with long, rather narrow leaves, and again with short and very wide leaves; the number of leaves on a whorl also being very variable, while in some plants the flowers are much more nodding than in others. In low, damp woods, near by where it is extremely abundant and attains a very large size, I have also noticed much variation in the shades and spots. The largest and most distinctly marked flowers I have seen occurred in strong plants having what might be termed variegated foliage, the leaves in these instances being yellowish in tone, marked with dark-green veins and blotches. Some have the back of the petals marked with pale-yellow bands on the edges. This is one of the most striking and exceptional forms, though the numerous flowers are smaller. Some have flowers with the under side of the petals stained a deep vermilion; some have large and some small dots; others occur with flowers much larger than the type; and the form I have specified as occurring with variegated foliage has the handsomest flowers of all, of medium size, with the back of the petals colored a glowing vermilion-scarlet. The large-flowered form has the petals the least spotted of all, no dots appearing on the terminal half of the petals. The latter is one of the most robust of the section. Another rare form occurs with the outside of the petals blotched and spotted with yellow, and I have met with still another form, intermediate between *rubrum* and *flavum*. All these, with the exception of the variegated form, I have found growing in the same woods in flickering shade, and all have preserved their distinguishing characteristics under cultivation. The yellow *Canadense*, while a less robust grower, withstands the sun better than the red variety. The latter is worthless grown in open sun. Placed among any of our native ferns except the big ostrich, which starts so early in growth as to choke or stunt the lilies, they thrive luxuriantly, and are thus probably seen to the best advantage.

Mr. Peter Henderson has justly remarked that the lily has no poor relations, and that in a general collection of the species all that can be imagined desirable and perfect in floral forms will be realized. Indeed, it is beautiful in all its very numerous forms; and when we consider that except one or two species it is a flower with no insect pests, the lily may well be regarded as one of the greatest treasures of the hardy flower-garden.

The roses seem more beautiful than ever this year, a characteristic of this favorite flower; it is always more beautiful. Said a blush rose to me: I am the type of youth and voluptuousness; I am red with the flush of health; with my odor, with my loveliness, all are intoxicated; I nestle in the bosom of beauty and I am the symbol of love; my beauty speaks for me. Do I need to trace my lineage? "I came of nectar spilled from heaven. Love, who bore the celestial vintage, tripped a wing and overset the vase; and the nectar, spilling on the valleys of the earth, bubbled up in roses."

There is so much to say about the rose that it were more satisfactory to recommend the reader to peruse the hundreds of monographs it has inspired than to attempt to allude to it within the confines of a few pages. The only way to do it justice is to begin at the beginning and treat it in all its phases of origin, history, culture, form, color, and fragrance. I imagine it would be delightful to study roses for a decade and then write a book. Even the subject of suitable manures would lose its taint if considered with reference to the rose. The species alone number upward of a hundred; the varieties with their briefest possible descriptions would fill a ponderous folio. Of teas alone there are several thousands; of hybrid perpetuals or remontants there is almost as great a multitude as the daffodils Wordsworth saw dancing by the shore of Ullswater. An astronomer it would require to count them; a Symonds to depict their colors. The rose, like the lily, will not grow itself, notwithstanding its hardy species are far less fastidious with regard to soil and climate. As the price of its beauty it requires care, if not "eternal vigilance." It is like a fascinating woman whom every one admires and who graciously submits to the attentions of all, to her own annoyance and discomfort. Thus, Madame de Coigny, becoming tired of the attentions bestowed upon her, one day had a signet engraved of a rose besieged by insects, with the motto—

This it is to be a Rose.

The first leaves have scarcely appeared ere they are beset by the thrip or rose-hopper, almost immediately succeeded by the green fly, leaf-roller, rose-chafer, and rose-slug. Were the sparrow of any earthly use, he would not leave these to hellebore, whale-oil soap, and Paris green. Nearly any one of these pests, if left to itself, soon ruins the foliage or flowers. Undoubtedly the easiest way to cultivate roses is to buy them; the next easiest way is to have a *posse* of gardeners whose sole purpose shall be to stand over them continually with wisp, bellows, and syringe. Indeed, it is far easier to study the lily and cajole its caprices than to escape the omnipresent thorn of the rose. There are roses without thorns as there is a bee without a sting; but a thornless rose nearly always means a rose without fragrance. But what loveliness it gives us to make up for its poutings—a dimple and a smile on every flower!

Who shall decide which rose is the type of beauty or say which is the sweetest? Can there be anything more beautiful than a Maréchal Niel? Is any rose finer than the combination of buff and peach-blow and salmon in the fragrant folds of a Gloire de Dijon? Is Louis Van Houtte or Marie Rady the sweeter flower, and are either of these as sweet as La France or Souvenir de la

Malmaison? And are these in turn as delicious as the little violet-scented white *Banksia* or the pungent breath of the white-clustered *multiflora?* Who shall choose between Marshall P. Wilder, Marie Beaumann, and Alfred Colomb? And which is the more bewitching, Madame Gabriel Luizet in her dress of pink Chambray, or Mabel Morrison in her cool white lawn? Which do you prefer, the old-fashioned climbers smothered in rosy bloom, or the mass of the Persian's beaten gold? Can you decide between a pink Bon Silene and a moss-rose wet with dew? Would you leave out the pæonia-flowered Paul Neyron for Madame the Countess of Serenye? And which is the more desirable in autumn, the colored hips of the dog-rose or the late-blossoming Marguerite de St. Amande? Then the white *Rosa rugosa*, the sweet-brier, the little Pacquerettes, the Noisettes, the Ayrshires, the Bourbons, the Chinas, the Boursaults, the damasks, the Provence, the Scotch, and the hosts of hybrids.

Which is my favorite in the hardy rose-garden? I have tried for many years to decide, and if pressed hard for an answer I think I should name Marie Rady, although not a few of the varieties I have specified and some I have not mentioned approach it very closely in the attributes which go to form a perfect rose. It is an ideal rose in form, color, fragrance, and foliage when well grown, perhaps not quite as free blooming as one might wish, and possibly more satisfactory as a budded plant than when grown on its own roots. I know of no rose more rose-like in its large, full, vermilion-crimson flower, its full, delicious perfume, its red-thorned shoots, and free, lustrous foliage.

But some like the brunettes and some the blondes. Both are beautiful, unless it be the type which loses its color with the first hot sun. Of course, there are many species which are not sufficiently hardy for the garden; but there still remain enough to puzzle any one to choose from. Some one has said that roses in a garden are preferable to a garden of roses, the latter at times affording little poetry or pleasure compared with a few roses here and there in a garden. An admirable plan, I think, is to plant enough of good forms and colors in the flower-borders; of Persians in the shrubberies; of climbers on the walls and pillars and trellises, and of all desirable hardy kinds in the kitchen-garden to cut from; and ever, and still ever, when wet with morning dew—

> Gather ye rose-buds while ye may,
> Young June is still a-flying.

Warm-Weather Wisdom.

Gods grant or withhold it, your "yea" and your "nay"
  Are immutable, heedless of outcry of ours:
But life *is* worth living, and here we would stay
  For a house full of books, and a garden of flowers.
          ANDREW LANG—BALLADE OF TRUE WISDOM.

VIII.

WARM-WEATHER WISDOM.

THE intense heat and long-continued dry weather are telling upon the flowers, and, at present, watering is the most important of garden tasks. Vainly have the hair-bird and tree-toad portended rain. It is one of the dry spells when all weather signs fail. The garden-hose, however persistently applied, only partially supplies the deficiency. The only thing that sounds cool.is the plaint of the mourning-dove from the depths of the thicket and the humming of bees in the lime. Even the swallows seem to fly less swiftly and the butterflies pass by less buoyantly. It is the sort of weather to reread the "Castle of Indolence" or the "Midsummer-Night's Dream." Some one should make out a list of books for reading during the reign of the dog-star. I should recommend, besides numerous volumes I have previously alluded to, such books as these as a sort of mental julep to sip when the thermometer is in the nineties: "The Unknown River," "The Life of the Fields," "I go a-Fishing," "Rambles among the Hills," "A Year among the Trees," "Walden," "Wind-Voices," "A History of Champagne." There is no end of cooling literary beverages, volumes that one can take up and skim through, Bulwer to the contrary notwithstanding, that reading without purpose is sauntering, not exercise—a winter rather than a summer maxim.

"The Haunted House" is cooling, and "In Memoriam" is nice to dive in. A fresh breeze blows perpetually from the "Penseroso"; "The Faerie Queen" is cool reading, and so is "The Plea of the Midsummer Fairies." All the noted sonnets on sleep are cool. Dobson or Lang ought to collect them in book-form between snow-white covers for hot-weather use. I remember a "Phantom Ship" (not Hamilton's sonnet) which used to provoke a cold shudder; but it is so long since, I have forgotten the authorship. There is also a "Phantom Fisher" somewhere in British verse—a spectral angler who draws ghostly trout from haunted shallows; and Whittier, besides the "Dead Ship of Harpswell"—"the ghost of what was once a ship"—has a phantom "Farm-House," wraith of a dead home. But, cooler than all of these, or any chill-provoking verse I recollect, is Tilton's "Phantom Ox," a rendition of the old Swabian superstition that a specter in the form of a white ox glides through the villages and farms, and that any person on whom he breathes at once sickens and dies. A little child, frightened in from his play, tells his mother, with blanched cheek and trembling lip, how, while wading along the brook in quest of lilies, a ghostly ox came down to drink. Through his body the trees, meadow-grass, and stones showed as through a crystal glass:

>He wandered round, and wherever he went
> He stepped with so light a tread,
>No harebell under his hoof was bent,
> No violet bowed its head.
>
>He cast no shadow upon the ground,
> No image upon the stream;
>His lowing was fainter than any sound
> That ever was heard in a dream.
>
>" I quivered and quaked in every limb!
> I knew not whither to flee;
>The further away I shrank from him,
> The nearer he came to me.
>
>" My handful of lilies he sniffed and smelt:
> His breath was chilly and fresh;

His horns, as they touched me softly, felt
    Like icicles to my flesh.

        .   .   .   .

" I rushed through the water across the brook,
    And high on the shelving shore
I stopped and ventured to turn and look,
    In hope to see him no more.

" He walked in my wake on the top of the flood,
    And followed me up the bank !
A blast from his nostrils froze my blood !
    My spirit within me sank !

" I hid in the reeds, O mother dear,
    But swift as a whiff of air
He followed me there ! he followed me here !—
    He follows me everywhere !

" Oh, frown at him, frighten him, drive him away !
    He's coming in at the door ! "
And down fell the lad in a swoon, and lay
    At his mother's feet on the floor.

The mother looked round her, dazed and dumb,
    She saw but the empty air,
Yet knew if the Phantom Ox had come,
    The shadow of Death was there.

She caught the pallid boy to her breast,
    And pillowed him on his bed ;
The white-eyed moon kept watch in the west ;
    The beautiful child lay dead ! *

---

* Theodore Tilton, " Swabian Stories."

This is as powerful as the "Erlking," and it deserves a place as a companion-piece to Schubert's grand rendition of the German lyric.

Doctors in summer should prescribe a light literary course, tonic rather than stimulating, not only to the weak-kneed, but the robust as well—on the same principle that salads, cooling vegetables, and dainty *entrées* are craved by the stomach during the tyranny of Sirius. I would further proscribe heating music : Strauss's waltzes, Von Weber's "Invitation to the Dance," Mendelssohn's "Wedding March," even Beethoven's "Adelaide," are entirely out of place during the heated term. Rather let us listen to the solemn chords of the "Dead March in Saul," the "Lacrymosa" of the "Requiem," the sobbing of the "Serenade."

The worst of existing hot-weather customs is that of sending bills in July. A law should be passed rendering this an indictable offense, if, withal, creditors should not be compelled to deduct a liberal percentage from all accounts falling due during the summer solstice.

Planchet's motto, "*Laissons faire et ne disons rien,*" is a good one for summer, and preferable to D'Artagnan's, "*Faisons bien et laissons dire.*" Happy in July is the man on the sea-shore ! How refreshing it is to get it all wet on one side of you, to have the ocean-breeze spraying you all the way in from the horizon, and to know the privilege of bathing with your lobster before eating him!

Under the lime-tree's shadow I find the coolest place of the garden. Is it due altogether to the shade, or partially to the myriad insect wings hovering unceasingly over the blooms above me? The ferns in the fernery near by look cool. Does a fern ever look otherwise than cool, and is not green always the coolest of colors? Cool are the lilac-scented white stars of the partridge-vine, almost covering its deep-green leaves. Cool, too, are the aspens on the hill-side which the wind visits when he passes by all other trees. And are not the tall, wild lilies cooled by their fluttering whorls? Despite their warm color, somehow their red Turk's-caps do not look warm, whereas the brick-red of the meadow-lily and the live coals of the scarlet martagon do in comparison. The wild lilies are now mostly in full vermilion bud and flower, some of them rising six feet high amid the ferns. The sight of their great candelabras of from six to a dozen flowers more than atones for the sting of the nettles and the labor of extracting their brittle rhizomes from the network of roots amid which they were entangled.

I thought the bouquet of the wild grape the most delicious breath of June; but now beneath the lime-tree's shade, lulled by the drowsy murmur of the bees, there seems no summer odor quite so fresh and uncloying as that of the blossoming lime. No wonder the honey probed from its scented cymes in the Lithuanian forests rivals that of Mount Hymettus thyme and is considered "the finest in the world."

The lime, a summer home of murmurous wings,

sings Tennyson. It is a very Mecca for the bees, and rivals its near neighbor, the Japanese honeysuckle, in the numbers of insects it attracts. What a motley throng of pilgrims are drawn to its nectar-laden shrine! Can it be the sweetness of its sap, which yields a sirup similar to the sugar-maple, that the ants and borers seek beneath its rind, eventually splitting the bark and destroying the tree? I believe this is peculiar to the European lime when grown in this country. De Gelien observes: "Many are fond of bees; I never knew any one who loved them indifferently—*on se passionne pour elles !*" The ancients were good bee-masters, in proof of which it may be cited that the Greeks had three terms at least for the different qualities of propolis or bee-gum—πρόπολις, κόμμωσις, and πισσόκηρος. The mead or metheglin of Shakespeare, the drink of the ancient Britons and Norsemen, and a favorite of Queen Bess, is very plausible, if not true, from the Greek, μέθυ αἰγλῆεν. Whoever is interested in bees will have read the fourth Georgic, and pondered the rules laid down by Butler. A better bear and bee story than that contained in "Reynard the Fox" is related by Butler, the *raconteur* being Demetrius, a Muscovite ambassador sent to Rome:

"A neighbor of mine," said he, "searching in the woods for honey, slipped down into a great hollow tree, and there sunk into a lake of honey up to the breast, where, when he had stuck fast two days, calling and crying out in vain for help (because nobody in the mean while came nigh that solitary place)—at length, when he was out of all hope of life, he was strangely delivered by means of a great bear, which, coming thither about the same business that he did, and smelling the honey (stirred with his striving), clambered up to the top of the tree, and thence began to let himself down backward into it. The man, bethinking himself, and knowing that the worst was but death (which in that place he was sure of), beclipt the bear fast with both his hands about the loins, and withal made an outcry as loud as he could. The bear, being thus suddenly affrighted (what with the handling and what with the noise), made up again with all speed possible; the man held and the bear pulled, until with main force he had drawn *Dun out of the mire;* and then, being let go, away he trots, more afeared than hurt, leaving the smeared swain in a joyful fear."

Scarcely less amusing is Butler's account of honey as a medicine, or his directions to avoid being stung by bees. They are as quaint as some of Walton's passages, or the directions by other old masters of the line for capturing a wary tenant of the stream. Walton has contributed one of the best *mots* that has appeared, on the frog: the instruction he gives Venator for baiting a hook with a live batrachian, which he commands him to use "as if he loved him, that he may live the longer." This is almost as realistic as another injunction by a Michael Angelo of the piscatory art, mentioned by Jesse, who would have a frog attached "to a goose's foot, in order to see, good halynge, whether the goose or the pyke shall have the better." Still another master of the antique school, speaking of the best bait for a pike, exclaims, with an enthusiasm for his art not to be met with in these degenerate days: "But the yellow frog, of all frogs, brings him to hand, for that's his dainty and select diet, wherein Nature has placed such magical charms that all his powers can never resist them, if fastened on the hook with that exactness, *that his life may shine*, and the bait seem undeprived of natural motion." When Theocritus sang, "Sweet is the life of frogs," he little thought of the pike, and the use the classic *Rana* would be put to by the modern angler. I think these old angling authors should be read during a midsummer drought—their stories are so cool, and ripple from their quills so spontaneously.

In connection with bees and insects, Jesse himself provokes a smile when he declares that, together with wasps and bumble-bees, the hornet "may be perfectly managed.... Two or three whiffs of tobacco-smoke, used as a fumigator, with a rose-nozzle—a very small one, that can be held between the teeth, is large enough—will instantly tranquillize all such insects, and render them quite harmless as to their sting; making them appear as if they had forgotten they possessed such formidable weapons.... The sting of a wasp is the least painful of all," he paradoxically continues; "the sting of a hornet I have never felt, nor that of the largest bumble-bee." But Jesse is not often caught napping, despite this paradox and his itinerant fumigator. It is, nevertheless, to be regretted that he thus deliberately denied himself the pleasure of a sensation which every one ought to experience at least once in a lifetime.

I consider Dr. Talmage a better authority than Jesse—he has felt the hornet's sting. I did not know him as an entomologist until he preached his sermon on "Stinging Annoyances," from the text, Deuteronomy vii, 20, "The Lord thy God will send the hornet." How vividly he describes him! "It is a species of wasp, swift in its motion and violent in its sting. Its touch is torture to man and beast. We have all seen the cattle run bellowing from the touch of its lancet. In boyhood we used to stand cautiously looking at the circular nest hung from the tree-branch, and, while we were looking at the wonderful pasteboard covering, we were struck with something that sent us shrieking away!"

The hornet is used as a simile for the stinging vexations of life which beset mankind in a thousand forms. If Talmage had a garden, he would see a swarm of hornets in the rose-pests, the dry weather, the overplus of rain, the plant-staking, the weeds, his dandelioned neighbors, the east wind, before which all plants must bow and many break. Indeed, he refers to the hornet as visiting us in the shape of friends and acquaintances who are always saying disagreeable things, and selects him as the type of the insectile annoyances of the world—these foes, too small to shoot, that are ever puncturing us one way or another. The Colorado beetle, the curculio, the locust, the Western grasshopper, the slug, the aphides, the currant-worm, the codling-moth, are all hornets in disguise. Perhaps the parson's solution, that the hornet is sent to "culture our patience," is the most rational one yet assigned for his existence. And yet the hornet is useful in another way, in feeding his young with the soft parts of other insects, including mosquitoes, which are thus largely destroyed.

The honey-bee is the most frequent among the insect visitors to the blossoms overhead, though the gnats and flies are also numerously present, banqueting on the sweets. I see various bumble-bees, wasps, and hornets as well, the former being the most numerous, after the honey-bees. From all of these many wings there arises a soothing, sonorous murmur of industry, a humming as from a vast hive. It is one of the sweetest of Nature's voices; less ethereal but not unlike the aërial music which one sometimes pauses to hear near woods and streams at this season. After Beethoven returned from wandering about a wood near Vienna, where he listened long to this aërial melody, he composed the grand Pastoral Symphony. This same sound puzzled the Selborne rector, in the Money-dells, over a century ago. Did this not also suggest the sound—

> That sometimes murmur'd overhead,
> And sometimes underground,

of Hood's " Elm-Tree "—Hood's lines being descriptive of the characteristic rising and falling of this woodland voice?

I remember hearing it repeatedly, years since, on still, hot days, in a small copse on a high elevation; and on revisiting the locality, recently, the same mysterious music followed me through the wood. Who are the performers of this gossamer-spun sound, this invisible harpsichord, this elfin music of the air? I have not seen a cause ascribed to it by the naturalists, though, it would seem, it must proceed from the trembling wings of myriads of midges, engaged in the dance of rivalry and love. Swinton's exhaustive volume on "Insect Variety," which treats so fully of the noises and dances of insects, throws no new light on the subject. Insects, and the swallows who pursue them, soar higher as the temperature becomes hotter; and it is, therefore, not improbable that the music produced by the fanning of innumerable wings should be distinguished when the performers are invisible.

I see a humming-bird visiting the wild lilies—he can not resist his favorite color—red. In a moment he darts to the lime-tree, but only for a moment, when he is rifling the blooms of the Japanese honeysuckle, where he remains suspended for a long period, often joined by the female. The boom of his swiftly-vibrating wings is audible where I sit; it seems as if they cooled the air! In the garden he skims rapidly over the borders, pausing a minute over the blue larkspurs, invariably visiting the scarlet lychnis and Chalcedonicum lilies; never neglecting the red monardas, and always returning to the honeysuckles. In Prof. Grant Allen's "Pleased with a Feather" I learn that the metallic luster of his topaz, emerald, and ruby-tinted throat is due to the fine lines of the feather barbules; and these it also is which give the sable sheen to the crow, whom I admire treading his favorite corn-field.

My Insect Visitors.

Ein Blumenglöckchen
  Vom Boden hervor
War früh gesprosset
  In lieblichem Flor;
Da kam ein Bienchen
  Und naschte fein:
Die müssen wohl beide
  Für einander sein.
        GÖTHE, GLEICH UND GLEICH.

IX.

MY INSECT VISITORS.

AS I listen to the humming of the bumble-bees, I think the term "bombination," formerly applied to the droning of the large *Bombus*, should be retained. It is expressive, and carries the sound shed by him of the black-velvet coat and winnowing wings. Sir John Lubbock's experiments with regard to the color-sense of bees are interesting as detailed in Chapter X in the volume "Ants, Bees, and Wasps." Repeated experiments made with honey placed on papers of different colors not only indicated a liking for blue on the part of bees, but showed a very decided preference for this over all other colors. Since the researches of Lubbock, Darwin, Wallace, and Müller, it is now well known that special colors in flowers are definitely designed to attract certain special kinds of insects; as, for instance, flowers that are intended for fertilization by various small flies are generally white, those which are designed to attract beetles are usually yellow, and those which depend upon bees and butterflies are almost always red, lilac, purple, or blue. Blue flowers, Prof. Allen observes, are, as a rule, specialized for fertilization by bees, and bees therefore prefer this color, while conversely the flowers have at the same time become blue because that was the color which the bees prefer. As in most other cases, the adaptation must have gone on *pari passu* on both sides. As the bee-flowers grew bluer, the bees must have grown fonder and fonder of blue; and as they grew fonder of blue, they must have more and more constantly preferred the bluest flowers.*

A singular preference of the large bumble-bee (*Xylocarpa Virginica*) has come under my notice in the case of the big bee-larkspur (*Delphinium Wheelerii*). One of the most robust and large-spiked varieties, I should not recommend it for the flower-border, both its small flower and peculiar color being less pleasing than numerous other varieties. It is growing side by side with handsomer and equally conspicuous kinds, and I should have discarded it

---

* Cornhill Magazine, "The Colors of Flowers."

long since were it not for the fascination it has always offered to the bumble-bees. The color of the sepals is a peculiar sky-blue, rayed with pale violet; the two spur-petals that project above the two yellow-bearded petals being dark brown and showing almost black against the contrasting color. Between these the bee ordinarily plunges his proboscis into the nectary; but the large black bumble-bee I refer to rarely if ever does this, but drives his spear into the spur of the flower from the outside, close to the base of the spur where the honey is stored. Perhaps this is done to save time and labor, or it may be owing to his short proboscis. He performs his work rapidly and assiduously, often remaining until stupefied from his banquet. This species has a habit of hovering over the flowers or in mid-air with a loud bombination, while chasing his mate, and seems more alert and quickly alarmed than others. When *D. Wheelerii* is in blossom little attention is paid to any other larkspur or any other flower of the garden, though numerous varieties of the bee-larkspur are far more odoriferous. The dark centers of the flowers are, of course, a conspicuous guide to the nectary; but similar centers exist in many other varieties.

So marked is the preference shown by this

species of bee for the variety specified, that on placing a large bunch composed of four other varieties of the bee-larkspur side by side with *Wheelerii* and experimenting with nearly a dozen different bees, in every case the insects when intercepted by the foreign flowers merely alighted on them for an instant, and, without inserting their proboscides, at once deserted them for the variety they were frequenting. I have been unable to determine whether it is the peculiar shade or some special odor of the flower which causes it to be sought out above the others, or whether it is on account of its being richer in honey. Besides this species I find other principal visitors in *Bombus terrestris* and the smaller bumble-bee, though none nearly as numerous as the large black species. *B. terrestris* also usually obtains his sweets from the outside of the nectary; but the smaller bumble-bee generally draws his nectar in the legitimate way. An occasional honey-bee searches for sweets through the aperture which has been made for him by a stronger lancet than his own.

With regard to the perforation of the corolla by bees, Darwin states that those plants, the fertilization of which actually depends on insects entering the flowers, will fail to produce seed when their nectar is stolen from the outside;

and even with those species which are capable of fertilizing themselves without any aid there can be no cross-fertilization, and this, as we know, is a serious evil in most cases.* Aristotle noticed that all kinds of bees and certain other insects usually visit the flowers of the same species as long as they can before going to another species, and it is a well-established fact, readily observable in any flower-garden, that bumble- and hive-bees will visit plants of the same species of opposite colors; but I am puzzled to account for the marked preference in the instance cited. The plants referred to are situated in a long row, and are considerably more numerous than any other variety. Yet this fact would hardly account for the preference I have noticed, for several seasons.

I think we do not accord the Germans sufficient credit for what they have accomplished by their painstaking and invaluable investigations in the interest of plant-knowledge. The ear-splitting terms they have to make use of and contend with! Just think of having to know that the "Sauerstoffabscheidung" of green plants is an "Ernährungsvorgang," and that the latter is closely connected with the "Lichtvermittelter

---

* Self- and Cross-Fertilization, chapter xi.

Desoxydationsprocess"! Is it any wonder it requires a "scholar of Trinity College, Cambridge," to translate a German scientist?

The first stimulus to more exact observation and distinction of plants was necessity—to know the countless medicinal species and to avoid confounding them with others. The old herb-gatherers were the first botanists. But since Aristotle, Theophrastus, and Dioscorides, who were herbalists rather than botanists, how much is the present system and knowledge of this science indebted to the Germans! What flowers have not been analyzed through their busy magnifying-glass — beginning with Brunfels and Fuchs; continued by Erhart, Hoffmann, Kömpler, Rumph, Hermann, Schreber, Sprengel, Göthe, Humboldt; and followed by Meisner, Endlicher, Meyen, Link, Schleiden, Von Mohl, Seubert, Müller, and others!

To all who would look beneath the surface and grasp the real purport and significance of flowers, Hermann Müller's volume, "The Fertilization of Flowers," recently translated into English, will be found of signal interest. Christian Sprengel, in 1787, was the pioneer to discover these fundamental truths:

1. The nectar of most flowers is secreted for the sake of insects, and is protected from

rain, that the insects may get it pure and undefiled.

2. The colors and odors of flowers are designed to attract the attention of insects.

3. Without the aid of insects very many flowers are incapable of fertilization, and therefore the secretion of honey in the flower, its protection, the odor of the flower, and the coloring of the corolla, are Nature's contrivances to cause its fertilization by insects.

While bringing forward the fact, however, that the pollen was conveyed by insects to the stigma, no greater advantage was assigned by Sprengel than direct contact of the reproductive organs—in itself no advantage over natural fertilization—without suspecting that the real value of insect-visits to the plant consisted in the pollen being thus carried to the stigmas of *other* flowers, and by this means accomplishing *cross-fertilization.* So, Sprengel's work, "The Secret of Nature in the Form and Fertilization of Flowers discovered," was allowed to lie fallow until called up again by the advance of knowledge and the researches of modern scientists, more particularly by Darwin's great work, "The Origin of Species," and his later book, "The Fertilization of Orchids." Müller's work, published much later than those of Darwin, besides the author's own marvelous researches and observations, includes references to everything of importance which had been written upon the subject prior to its publication. Of the mass of information here presented, the enumeration of the various flower species, with their throngs of visitors, is one of the most noteworthy features.

It will prove interesting, perhaps, to recapitulate briefly the forms and character of insectivorous life which serve to carry on the process of cross-fertilization. "A review of the mode of life of insects which visit flowers, and of the families to which they belong," says Prof. Müller, "shows continuous gradations from those which never visit flowers to those which seek them as a secondary matter, and finally to those which entirely depend upon them. This shows clearly that insects which originally did not avail themselves of flowers gradually became more and more habituated to a floral diet, and only became correspondingly modified in structure when they had learned to depend upon such a diet exclusively."

In the scale of importance as fertilizers, the order of *Hymenoptera*, to which belong the bees, takes the highest rank, its members in the perfect state being entirely dependent on flowers. Bees, which confine themselves exclusively to a floral diet, have led to more adaptive modifications in these flowers than the *Orthoptera* and *Neuroptera*, the *Hemiptera*, the *Coleoptera*, and the *Diptera* and *Thysanoptera* combined. To them we owe the most varied, most numerous, and most specialized forms, the flowers adapted to the *Apidæ* probably surpassing all others together in color-variety.

The *Hemiptera*, to which belong the bugs, stand higher than the *Orthoptera* and *Neuroptera*, to which belong the cockroaches and dragon-flies, several species being fitted by their small size to creep into and suck honey from very various flowers. The *Coleoptera*, to which belong the beetles, are of much greater importance as fertilizers, for many species in widely different families feed at times on flowers, and a still greater number confine themselves to such food exclusively. On the other hand, the voracious beetle does much harm to numerous flowers by nibbling their reproductive organs.

The *Diptera*, to which belong the flies and gnats, stand on a still higher plane than the *Coleoptera* in the matter of adaptation to a floral diet, and are of far more importance for fertilization, the majority of *Diptera* resorting to flowers. In the habits of the *Empidæ* of the general order *Diptera*, Müller clearly sees the transition from blood-sucking to honey-sucking. Sometimes in a single species the females, which require more nourishment, are blood-suckers, while the males seek honey only. In *Paltostoma torrentium* (*Blepharoceridæ*), two different kinds of females exist together, one blood-sucking, the other honey-sucking; while the males are all alike, and all feed on honey. In like manner, Müller states that several flowers seem to have acquired an offensive smell correlative to the habits of certain anthophilous flies which at times feed on putrid flesh and excrement as well as flowers. Tiny species of midges, which people dark corners by day and leave them in the evening, are regular fertilizers of many flowers which afford somber hiding-places for their visitors.

In almost all bees highly specialized for fertilization, the body is more or less thickly clothed with long, feathery hairs, that in many flowers become dusted, without any direct effort, with a considerable quantity of pollen, which is then cleared off by means of the tarsal brushes. Easily as the hairs take up pollen, they return it with equal ease to viscid or rough stigmas.

So greatly has the hirsute covering of the hind-legs increased, and so perfect has become the development of tarsal brushes in the exceedingly numerous species of *Halictus* and *Andrena*, that the practice of feeding the young on pollen collected by these hairs is exclusively or mainly relied upon. In all species which provide for their own young, the males are of far less service for fertilizing plants than the females, as they are merely interested in their own maintenance, and neither collect pollen nor visit flowers very diligently. Yet, in all species in which a more or less thick coat of feathery hairs has become developed upon the bodies of the females, it has become transmitted by inheritance to the males, so that they also serve as pollen-collectors.

Think of the number of bees alone that take part in the process of fertilization!—bees with abdominal collecting-brushes and long proboscides; the specially long-tongued *Bombus* and *Anthophora;* other bees with long or moderately long proboscides; bees of the genus *Prosopis*, themselves possessing a peculiar odor, and preferring highly odorous flowers; *Andrenadæ* and *Apidæ;* hive-bees and bumble-bees; workers and drones; big bees and little bees; and almost every variety of *Hymenoptera* with a sting in its tail.

The *Lepidoptera*, to which belong the butterflies and moths, are likewise highly important agents in the evolution of flowers, for which they are peculiarly fitted by their long, thin proboscides, enabling them to probe the most various flowers, whether flat, long, or tubular. Even at night, in fragrant gardens, in lonely meadows, in the most sequestered woods, the process of insect fertilization goes on continuously. Then it is that the great nocturnal hawk-moths, their two immensely long, hollow laminæ coiled in a spiral, emerge at twilight to haunt the lighter-colored flowers, which exhale their odor most powerfully at night. Verbenas and petunias, always intensely fragrant at this time, are especially sought out by the crepuscular-*Lepidoptera*. Like the humming-bird and swallow, the body of the great sphinges, tapering at the tail, and the stiff, pointed, sharply-cut wings, are framed with special reference to agility and sustained flight—agility to avoid their pursuers, and great strength of wing to sustain constant suspension in mid-air. I have seen the deliciously scented Japanese honeysuckle (*Lonicera Halleana*), on warm June and July evenings, swarming with the large *Sphingæ*, including *S. Carolina*, *S. cinnerea*, and the smaller *S. drupiferarum*, the former being present most numerously. The humming of the rapidly vibrating wings, the quick, furtive flight, the perpetual hovering over the flower-chalices, the curious coiling and uncoiling of the great suctorial tubes, are a feature in the strange processes of Nature that, once seen, can scarcely be forgotten.

Müller, from whom I have already drawn largely on this fascinating subject, referring to the dusk-loving *Lepidoptera*, explains that the rapid movements always characteristic of this species may be due "to the shortness of the period suitable for their flight, or *to the pursuit of bats.*" In cases which have come under my own observation, I have noticed that the *Sphingæ* appear most numerously at dusk, haunting their favorite flowers with little diminution for about two hours, and apparently decreasing in numbers as the night advances. Bats, I have repeatedly noticed, seem most abundant during the early hours of night. Tennyson's passage in " Mariana "—

> After the flitting of the bats,
> When thickest dark did trance the sky—

would corroborate this, if the poet meant to italicize the anterior preposition.

While angling for speckled trout at night in summer, I have observed, where the bats were very numerous, their sudden departure and subsequent appearance, at perhaps quite long intervals. Attracted by the abundance of stream-frequenting *Lepidoptera*, the *Cheiroptera* discontinued their aërial chase so soon as the quarry withdrew from the neighborhood of the water, returning with the reappearance of their prey. Upon the retirement of the bats, the trout in like manner ceased to rise freely to the artificial white moth, the time of the "take" being apparently regulated by the presence of the bats, though governed really, of course, by the return of the insects in their merry-go-round over the surface of the stream. Finally, the flowering period of my Japanese honeysuckle I have always found a certain index to the commencement of night-fishing.

I am not aware whether the great sphinx is too big a mouthful for the bat; he is certainly a *bonne bouche* for the greater and the lesser owls. Several summers ago I was awakened suddenly, about midnight, by a strange noise, as of some one raising the awning and tiptoeing on top of the veranda. A bright moon was shining, and not a breath of air was stirring. On the veranda's rim, looking down upon the honeysuckles and their honey-seeking visitors, stood two small screech-owls; while, startled from his perch upon the awning, a great horned owl flew away without a sound.

"The swift, violent movements of the *Lepidoptera*," the author of "Die Befrüchtung der Blumen" continues, "is of very great importance to the plants they visit ; for the more flowers that will be visited in a given time, the less the time spent on each, and the shorter the time spent in the flight from one to another. This explains how many flowers have adapted themselves specially to nocturnal insects, both by their light colors visible in the dusk, and by their time of opening, of secreting honey, or of emitting their odor. The *Sphingidæ* perform their work as fertilizers with peculiar rapidity, dropping their long proboscides into a flower while hovering over it, and instantly hastening away on their violent flight to another. Accordingly, most nocturnal flowers have adapted themselves specially to these *Lepidoptera*, hiding their honey in such deep tubes or spurs that it is only accessible to the *Sphingidæ*."

To the *Lepidoptera* is assigned the second or third place as fertilizers before or after the flies. No special mention of the humming-birds as flower-fertilizers is made by Müller, who confines his observations strictly to insects.

Very many flowers that are only accessible to the butterfly, moth, and humming-bird, on account of their long, contracted nectaries, have become, as we have seen, gradually developed or modified through the agency of their visitors; while the colors, odors, and periods of opening of flowers generally are in relation to the tastes and habits of the insects that frequent them. Odor, no less than conspicuousness, is a powerful magnet to the insect tribes; indeed, strong scent is even a greater attraction than brilliant colors.

Many flowers are both conspicuous and odoriferous. On this point Darwin observes: "Of all colors, white is the prevailing one; and of white flowers a considerably larger proportion smell sweetly than of any other color, namely, 14.6 per cent; of red, only 8.2 per cent are odoriferous. The fact of a large proportion of white flowers smelling sweetly may depend in part on those which are fertilized by moths requiring the double aid of conspicuousness in the dusk and of odor. So great is the economy of Nature, that most flowers which are fertilized by crepuscular or nocturnal insects emit their odor chiefly or exclusively in the evening. Some flowers, however, which are highly odoriferous depend solely on this quality for their fertilization, such as the night-flowering stock (*Hesperis*) and some species of *Daphne;* and these present the rare case of flowers which are fertilized by insects being obscurely colored."*

Thus we see how important a part the insect world has taken in the evolution of the floral world, and how much the beauty and variety of the garden owe to the myriad murmuring wings which ceaselessly ply their appointed task of development and improvement.

---

* " Cross- and Self-Fertilization," p. 374.

Hardy Shrubs and Climbers.

AMONG the links between man's mind and Nature we may place, as one of the most obvious, man's earliest attempt to select and group from her scattered varieties of form that which—at once a poem and a picture—forms, as it were, the decorated border-land between man's home and Nature's measureless domains, *The Garden*.—BULWER, MOTIVE POWER.

X.

HARDY SHRUBS AND CLIMBERS.

INCE the lilacs were in bloom there has been no lack of other shrubs to extend the blossoming season. The slender-branched *Deutzia* (*D. gracilis*), the rough-leaved (*D. scabra*), *D.* Pride of Rochester, *D. crenata fl. pl.*, and others of the species, are all indispensable flowering shrubs, appearing in the order named. Scarcely less effective are many of the shrubby *Spiræas*, which flower in the following order, from the middle of May to the middle of August: (1) *Prunifolia fl. pl., Thunbergii;* (2) *Niconderti;* (3) *chamædrifolia;* (4) *cratægifolia, lanceolata, lanceolata fl. pl., lanceolata robusta;* (5) *ulmifolia;* (6) *opulifolia aurea, crenata;* (7) *Fontenaysii, salicifolia, sorbifolia;* (8) *Billardi;* (9) *ariæfolia;* (10) *callosa, callosa alba, callosa superba,* species *Japonica.* *S.*

opulifolia aurea is a valuable variety, with golden-yellow foliage. It deserves a place by itself, or plenty of room to develop in the shrubbery. Soon after its blossoming period, the four-cleft seed-pods of the cymes, which are thickly clustered along its drooping branches, turn to a rich terra-cotta shade. This shrub, when once pruned into symmetrical shape, should receive as little pruning thereafter as possible, or the light, graceful effect of the sprays will be destroyed. *S. Japonica* is of recent introduction, a graceful, medium-sized species, covered in July with attractive, rosy-red cymes.

The native white fringe (*Chionanthus Virginica*), though classed as a tree, should be included among flowering shrubs. It is distinct and beautiful, with its glossy leaves and feathery plumes of pure white, fragrant flowers. Its odor reminds one somewhat of the native yellow-wood (*Cladrastis tinctoria* or *Virgilia lutea*), one of the finest of ornamental trees, with wistaria-like racemes of fragrant white flowers, profusely produced during midsummer every other year. This vies with the lime-tree in the attractions it presents to the bees.

Of the snow-balls there are a score of varieties to choose from. Among these, the common Guelder rose (*Viburnum opulus sterilis*) is among the best. The Chinese *V. plicatum* is the finest of all, surpassing the common variety in habit, foliage, and flowers, and it is not unjustly considered one of the most valuable ornamental shrubs. There are likewise very many varieties of the syringa, or mock-orange, to select from; some earlier and later, some with larger and smaller flowers, some odorless and some intensely perfumed. Doubtless the familiar garland syringa (*Philadelphicus coronarius*), one of the earliest to appear, is as satisfactory as any of the strongly scented kinds. *P. Gordonianus*, a late bloomer and vigorous grower, is more delicately perfumed. The golden-leaved syringa (*P. foliis aureis*) should be in every choice collection of shrubs; its shape is easily maintained, and its vivid golden foliage is valuable for enlivening the shrubbery or contrasting with purple-leaved subjects.

Some like the odor of the elder-flower; many do not share the preference shown for it by the flies. In any event, the variegated-leaved variety (*Sambucus variegata*), its foliage mottled with white and yellow, is one of the best variegated-leaved shrubs. The elder's cymes, produced so profusely, are always beautiful. The golden-leaved variety has vivid yellow foliage, but, somehow, appears to have a sickly look, or to have assumed an autumnal hue before its time. The fern-leaved and cut-leaved varieties are both fine; and *S. nigra*, a medium-sized European species, gives us handsome purple-black berries in the fall. The *Halesia*, or silver-bell (*H. tetraptera*), a beautiful, large shrub, with white, bell-shaped flowers in May, should not be overlooked. Besides its peculiar flower, it is distinguished by its large, four-winged fruit.

During latter June the laurel-leaved privet (*Ligustrum laurifolium*) is laden with its spikes of creamy-white flowers. This and the box-leaved variety are probably the two finest; the latter retains its thick, dark-green leaves for a long period, and both are of erect and handsome habit. But the privet is liable to suffer from extreme cold, and is slow to recover when affected. The white alder, or sweet pepper-bush (*Clethra alnifolia*), should have a partially shaded and sheltered position, as well as abundance of moisture, to do it justice, its natural habitat being swamps and low woods. A drive through the woods on the New England coast in August is rendered doubly delightful by its delicious breath, rising from the shaded thickets where it grows in the greatest luxuriance. On account of its graceful and fragrant flower-spikes, its neat habit, and its florescence when most other shrubs have passed, it should be seen much more frequently in the choice shrubbery. The button-bush (*Cephalanthus occidentalis*), which often keeps the *Clethra* company, is a desirable shrub, with attractive foliage, and round heads of sweet-scented white flowers appearing at the same period.

Kalmia latifolia, a member of the ornamental heath tribe, would be one of the most desirable medium-sized shrubs if it could be grown as it grows itself. But I find it useless to attempt in western New York, where artificial culture is entirely unsuited to it, under whatever conditions of soil and position it may be tried. The leaves of the *Kalmia* are said to be poisonous to some animals, and the honey derived from it has been known to prove fatal in several instances. It is always well to try new plants which one admires, or plants that have been recommended; but, when one does not aspire to having a botanical collection, it is also well to drop all subjects that one does not admire, or that prove themselves ill adapted to the climate. Still, a plant may be well worth cultivating in one climate and worthless in another—I might almost say, successful in one's neighbor's garden and a failure in your own, or *vice versa*.

Only through experimenting, however, can one determine what to attempt and what to avoid.

Of the *Diervillas*, or *Weigelas*, many of the so-termed rose-colored kinds, I think, are to be avoided. The nurserymen's catalogues swarm with the many varieties of this shrub. The typical color—"rose"—is poor, and I should condemn the *Weigela* as a garden shrub were it limited to its commoner form. The white varieties, on the contrary, are desirable, and so are some of the dark reds, which are not frequently seen. A clear, rose-colored variety, recently introduced under the name of "Othello," is an exception to the typical rose-color, and is possessed of much merit. The dark form, "Jean Mace," lately sent out, is distinct, its long, tubular, maroon flowers being specially striking in the bud stage. "Edouard André" and "Lavallei" are among the best of the dark hybrids, but the latter has a straggling habit. Most of the *Weigelas* are apt to grow straggling with age—an objectionable feature of the genus.

The *Hibiscus*, althæa, or rose of Sharon, is a charming adjunct to the shrubbery—neat in form, free-flowering, and always gay during late summer and September, when the shrubbery begins to look dull, and the sad-voiced crickets remind one that the floral beauty of the year has begun to wane. Both the double and single forms are fine; and the white and flesh-tints, with their distinct dark eyes, are the most pleasing colors. The purples and violet-reds are for the most part objectionable. "Painted Lady," the name of one of the varieties, well describes the lively flower of the althæa. The variegated-leaved variety is one of the finest variegated-leaved shrubs.

Hydrangea paniculata grandiflora, the great-flowered hydrangea, is a splendid late-flowering shrub, with its immense panicles and changeable shades, and it should enliven every garden in September. Nor should the early white-flowering *Exochorda*, the fragrant white upright honeysuckles, the sweet-scented *Calycanthus*, and the *Colutea*, or bladder-senna, attractive for its reddish seed-pods, be overlooked in the collection of shrubs.

Besides the flowering species, there are many shrubs which deserve a place on account of peculiar habit, characteristic foliage, or colored fruit. Of shrubs with dark-colored foliage, the purple-leaved barberry, purple-leaved plum, and many of the dwarf Japanese maples, may be specified. Of shrubs with variegated foliage, there are several varieties of the shrubby dogwood; several of the *Weigelas*; the silver-leaved *Corchorus;* the white-edged and golden privets; the golden syringa; the variegated-leaved elders; the variegated St. Peter's-wort; the variegated althæa. Numerous shrubs, also, are valuable for their ornamental fruit, which succeeds the flowers. In this class the following are all excellent: The common barberry, with scarlet and violet fruit in September; the red dogwood, with white berries in September; the red- and the white-fruited *Euonymus;* the red-fruited *Cotoneaster;* the Cornelian cherry, with its large and showy red fruit in August; *Elæagnus edulis,* with red, cherry-like fruit in midsummer; the red-berried *Viburnum opulus* and black-berried *lantanoides;* the black-fruited elder; and the snowberry. There are, moreover, many trees and shrubs, beautiful for their autumnal coloring, which should be remembered; these will be referred to in a subsequent chapter.

No garden is complete—if a garden can ever be complete—without its flowering climbers. Even the kitchen-garden should have its scarlet pole-beans, and the front veranda, at least, be festooned with blossoming vines. But there are so many desirable sorts, that all suitable places about the house and grounds should be utilized, to enjoy as many of them as possible. The wistaria alone holds a whole summer of fragrance in its June cascade of bloom. Those who care for variety have a number of kinds to choose from, though none equals the robust, hardy, and free-flowering Chinese blue. It is at home in any exposure, and only needs support to a sufficient height to prove one of the finest ornaments of the garden. By planting it on the north and on the south side of the house, its flowering period may be greatly extended, a vine placed in the former position coming into bloom just as one in a southern exposition is passing.

The numerous species and varieties of virgin's-bower, or *Clematis*, are beautiful for veranda and trellis decoration, as well as for fence-screens, for pillars along garden-walks, and for training on walls and arbors. Few hardy plants afford such combined beauty, luxuriance, and continuous bloom. For a full description of its hundred species and varieties, the reader should consult Moore and Jackman's "Clematis as a Garden Flower," the most comprehensive treatise on the subject. Of the several types, the *Jackmanni* and *Viticella* are the most generally seen—the common *Jackmanni*, all things considered, being the most satisfactory representative of the genus; these flower during summer and autumn in continuous masses on summer. shoots. The *Lanuginosa* type, of which the white *C. Henryi* is the finest example, flowers during the summer and autumn successionally on short lateral summer shoots; flowers dispersed. The *Viticella* type, represented by *C. v. venosa*, *C. v. modesta*, etc., blossoms in the summer and autumn, successionally, in masses, on summer shoots. The *Graveolens* type, flowering on the young growing summer wood, some of which are odorous, comprises a series of hardy, fast-growing species of easy culture. The *Montana*, *Patens*, and *Florida* types blossom on the old wood, and include the earliest or spring-flowering divisions of the family. The *Patens* type has supplied a large number of varieties, some of which are sweet-scented. To this section belong the fine, large varieties, Edith Jackson, Fair Rosamond, Miss Bateman, and others. In whatever form or color it occurs, whether appearing in sheets of purple, like *Jackmanni* or Alexander, or wreathing a roadside hedge with white garlands, like our native virgin's-bower, the clematis is a flower which always claims our admiration.

Most of the clematis are easily grown in rich, deep, friable loam, and should be mulched with old manure in winter, and given abundance of water during dry weather. Partial shade serves to develop the color and size of the flowers. English growers advise that the clematis be richly manured; some American growers, that it must not have manure about the roots. In any event, the best results are obtained by planting it in new soil, in partial shade. Not unfrequently the roots of many of the clematis become infested with a grub, which forms knobs along the fleshy roots, often impairing the vitality of the plant. A species of blight also attacks it sometimes, causing the plant to die down, but apparently not injuring it below the surface. The clematis is of comparatively recent introduction to this country, but has already become, through one of its types, at least—*Jackmanni*—the most popular, perhaps, of climbing flowering plants.

It is well that no one flower combines every quality, and that the more conspicuous forms of the clematis are odorless. Were this not the case, it would be planted still more extensively, and we should lose much of the variety which other climbers contribute. Showy as it is, it can not take the place of the climbing rose, the joy of many an arbor and veranda; or the honeysuckle, sweet as its mellifluous name. The latter is an old favorite, and one that no other flowering vine can well surpass. Attractive in all its forms, the recently introduced Japanese species, *Lonicera Halleana*, is its most beautiful representative for the veranda, arbor, trellis, or wall. This, though comparatively little known in Europe, is widely disseminated with us, where it was first introduced with the beautiful Japanese *Magnolia Halleana*, by Dr. Hall, of Elmira. I know of no climber that combines so many good qualities; for, independent of its vigorous growth and prodigality of fragrant white blossoms, it would be beautiful for its dark-green evergreen foliage, which it retains during a great portion of the winter.

With honeysuckles, as with many other things, however, absolute perfection is seldom found in a single variety or individual. While this species is as yet without insect-enemies, and is unquestionably hardy, it is nevertheless liable, even after having passed unscathed through several severe winters, to die down suddenly in spring, apparently from the effects of the cold. This is the case mostly with old plants, and I am not certain whether it is entirely a matter of climate, or whether it is not due partly to its habit of twining so closely as to strangle itself. But it is so rapid a grower that plants are soon replaced, and its odor is so delicious and its blooming period so continuous, that it is worth having at any pains. Its fragrant white flowers, passing to yellow, are, as previously stated, a powerful magnet to the bees and honey-seeking insects. The green leaf-cricket loves its shady tangle, and I always hear his first ode to autumn among its leaves. The long spur, recurved petals, and feathery stamens, characteristic of the honeysuckle, are seen to advantage in numerous other species, the next best to *Halleana* being the monthly fragrant, or Dutch, a vigorous climber, with red and yellow fragrant flowers appearing all summer. The Japanese golden-leaved variety is handsome, with its foliage netted or variegated with yellow.

For pillars and arbors, the native trumpet-flower (*Tecoma radicans*), and its darker form, *T. r. var. atrosanguinea*, are valuable climbing shrubs, becoming picturesque with age. The large, vivid red, tubular flowers in clusters are very conspicuous, showing well from a distance, where it may be placed to the best advantage as a pillar-plant. *Actinidia polygamia* is a handsome Japanese climbing plant, with dark, clean foliage resembling that of the apricot. Its flowers are white, with a purple center, and sometimes cover the whole vine, the fruit being round, edible, and fine-flavored. Other hardy flowering climbers which may be specified are

Akebia quinata, a singular Japanese climbing shrub, with fine foliage, purple flowers, and ornamental fruit; the large-leaved native climbing staff-tree (*Celastrus scandens*), with yellow flowers and clusters of orange-capsuled fruit; the moonseed (*Menispermum Canadense*), a native, twining, slender-branched shrub, bearing small yellow flowers and black berries; the silk-vine (*Periploca Græca*), a handsome, fast-growing European climber, with glossy foliage and purple-brown axillary clusters of flowers. The native wild bean (*Apios tuberosa*) is a small-growing vine, with racemes of reddish-brown, fragrant flowers, recalling the perfume of violets, which is best left to twine around the royal fern, with which it is almost always found in company.

Finally, among the native clematis, the common virgin's-bower (*C. Virginiana*), the fragrant *C. crispa*, and *flammula*, as well as *C. integrifolia* and *Pitcheri*, may each and all be used to advantage in the adornment of the garden.

In and out of the Garden.

The summer's flower is to the summer sweet,
Though to itself it only live and die.
                    Sonnet XCIV.

Not only the days but life itself lengthens in summer. I would spread abroad my arms and gather more of it to me could I do so.
    Richard Jefferies, The Life of the Fields.

XI.

IN AND OUT OF THE GARDEN.

MUCH-NEEDED rain has come at last—a steady, drenching, searching rain; a freshening, quickening, revivifying rain—a rain that has oozed down to the bottom, that has loosened the soil and cleansed the foliage, and sought out every root and rootlet beneath the ground. Light showers are of little service in time of drought; they are like the efforts of the garden-hose, and have no lasting effect. The colors of the flowers have come out with renewed intensity, and there is a marked increase in the luster of the foliage. The larkspurs are as brilliantly blue as the sky above them, and the scarlet lychnis (*L. chalcedonica*) burns as intensely as the setting sun. This is one of the most dazzling summer flowers; a single bloom of it, when well grown, will show its color to advantage. It is one of those peren-

nials that may be dotted here and there in the border; its scarlet is so strong, it does not require to be planted in masses. This varies somewhat in the size of its flower-heads and the intensity of its hue according to the soil and the season.

The same observation holds good with very many plants, that some years prove extremely satisfactory and again are disappointing. A thorough rain at the right time does wonders toward heightening the hues of flowers; and cool weather is everything in holding the true colors of many. Certain varieties of roses which faded rapidly one season, and which I had condemned on that account, I have found, another season, when the weather was favorable, entirely satisfactory. So that it is not always possible to judge of the merits of a flower from a single season's experience. Neither can one expect that a species which is desirable in one place will invariably prove so in another—so much depends on climate, soil, and the caprice of the weather.

A fine contrast to the scarlet lychnis, besides the larkspurs, roses, excelsum and candidum lilies, is *Chrysanthemum maximum*, a grand, hardy Marguerite, which has large white daisy-like flowers, with yellow centers, on stiff stalks. The narrow notched leaves are of a deep green, the foliage abundant, and the plant of elegant habit.

Bupthalmum cordifolium, the European ox-eye, is a stout perennial with large leaves, that opens its yellow blossoms the latter part of June, soon after *Anthemis tinctoria*. It is far too coarse to take the place of *Coreopsis lanceolata*, and is most suitable for the rear, or the wild garden.

Many of the *Centaureas*, the plant which cured the foot of Chiron, wounded by the arrow of Hercules, are valuable border-plants. The large blue flowers of *C. montana* appear early in June. This is not so neat in habit as some; but its blue is beautiful and the flowers charming in the cut stage. The flower of *C. Ruthentica* appears on a very tall stalk, rising high above the somewhat sparse foliage, shortly after the appearance of *C. montana*. The single blooms are large, but they only hold their color and freshness for a day or two. *C. macrocephala* is a robust, thick-foliaged species, with large bright-yellow flower-heads; and, while showy as a border-plant, it is not as fine as *C. glastifolia*, a more elegant plant, which succeeds it. I think this the finest of the large species, crowned in July with a perfect mass of golden bloom on branched stalks four to five feet high. The silvery buds themselves are handsome for several weeks before they open. *C. dealbata*, an earlier species from the Caucasus, is a medium-sized plant, with silvery foliage and pretty rose-purple flowers. The Persian sweet-sultan (*C. moschata*), though an annual, is always worth the trouble of growing.

One of the largest-leaved perennials is the great groundsel (*Senecio macrophylla*), the leaves of which attain an immense size in shade, but as yet I have been unable to cause it to flower; the leaves wither quickly in the sun, and it is also very sensitive to dry weather. *S. pulcher*, a very late species, bears large purplish blossoms, with yellow centers, a handsome and distinct flower, the best of its tribe. *Scabiosa Caucasica* is by far the best of its section of the teasel family, and, being a perennial, is more valuable than the biennial *S. atropurpurea*, also a handsome flower. When grown in congenial soil the former is a beautiful medium-sized border-plant, its large, flat lavender flowers being very distinct, and gracefully placed on tall stems.

In specifying *Lychnis chalcedonica* as one of the most dazzling reds, I meant no reflection on the scarlet avens or *Geum*. It has as bright an eye as a rabbit; at least, it is as red as a rabbit's eye. An inhabitant of the Bithynian Mount

Olympus, its single is beautiful, and its double doubly so, as it remains so much longer in perfection. The Japanese *Veronica longifolia subsessilis*, a midsummer flower of recent introduction, is unquestionably the finest herbaceous speedwell. Its flower is a lovely deep blue, and its foliage handsome. It is in all respects a superior border-plant; this species, however, does not make seed.

Where the climate suits it, the large horsemint (*Monarda didyma*), one of the best of the big labiates and the finest of the genus, is a valuable garden-flower. The leaves possess a strong mint-like odor, and the dark red of its flowers is striking. It is apt to encroach upon its neighbors, however, and requires abundant moisture. This species is said to give a decoction but little inferior to the true tea, and was formerly largely used as a substitute in Pennsylvania.

There are numerous species of the *Statice*, or sea-lavender, the best of which is *S. latifolia*. The *Statice* is invaluable for bouquets, and should be in every garden for cutting, to employ in the old-fashioned nosegay. One sometimes becomes tired of the regulation bouquet, composed of a single flower, and then the *Statice* helps one out. I see it now, its feathery sprays rising above the sweet-smelling nosegay composed of carnations, mignonette, feverfew, bachelor-buttons, Iceland poppies, pinks, larkspurs, sweet-williams, and lemon-verbena. There should always be plenty of these old-fashioned flowers to cut from.

The grand inflorescence of the chestnut-trees on the hill-side is mostly past—not, however, before the cicada rings out his song of heat. I invariably hear his first overture while the chestnut is still in bloom. I love his magnificent *crescendo*. How broad his diapason, and how sonorous the mighty volume of sound! It is the most fervid of all summer sounds, this ringing expression of drought and heat, produced by the hind-legs with which he leaps, said Aristotle two thousand years ago. It is pleasant to know, according to another classic—Zenachus—that the cicadæ live happily, since they all have voiceless wives; the two drums on either side of the body under the wings not existing in the female. The cicada's song brings up Meleager and Theocritus, the classic cicada, I believe, being a species of *Tettix* or harvest-fly, erroneously termed "locust." Independent of entomological accuracy, cicada is the preferable name; it has a drier and more sibilant sound. Virgil's cicadæ are *querulæ* and *raucæ;* Martial's, *argutæ* and *inhumanæ.* In the "Anthologia," on the other hand, they are always sweet singers. Meleager's cicada is a

> Charmer of longing—counselor of sleep!
> — The corn-field's chorister
> Whose wings to music whir.

Theocritus can only find in the cicada a minstrel sweet enough to compare with the song of Thyrsis:

> For sweeter, shepherd, is thy charming song,
> Than ev'n cicadas sing the boughs among.

There is much of the delightful old Hellenic philosophy in Thoreau's sentence: "The things immediate to be done are very trivial; I could postpone them all to hear this locust's song." I find the cicada somewhat like the rain—there is always an interval between the first drops and the down-pour, as there is between the first warning of the *Tettix* and his subsequent chorus of heat.

The grasshopper and cricket have but just begun their song in faint, quavering notes, which they will increase with the advance of the season, and the male green leaf-cricket is voiceless as yet on the honeysuckle-vine. These will atone ere long for the silence of the birds whose voices fail as the insect stridulation gathers force.

On sandy banks the butterfly-weed (*Asclepias*

tuberosa) was gay a fortnight since with orange corymbs. It is among the brightest of summer flowers and the most brilliant of the extensive milkweed tribe that crowds and perfumes the waste places during summer. Leaving the sandy places where it grows, I find the wild rose still in blossom How full the aroma held by its few single pink petals—a freshness and pungency its cultivated sisters do not possess for all their double cups and titled names! In the swamp further on, where virgin's-bower and purple nightshade wreath their festoons, there streams a veritable sunset of color. The gorgeous cardinal-flower (*Lobelia cardinalis*) is in full panoply of bloom—the most vivid red of the year, a red that seems endowed with conscious life, so glowing is its fire. Growing near it I find the great blue lobelia (*L. syphilitica*), a conspicuous flower, and more rarely its white form, with an occasional plant of the fragrant snake-head (*Chelone glabra*).

Something fascinating there is about a swamp—its rare flora, its gloom in daylight, its freshness in drought, its ever-present mystery. You can not grasp it as you can the dry woodland. The very birds are evasive, and its flora leads one deeper and deeper into the tangle where the woodcock springs from the thickets of jewel-weed and the owl skims noiselessly from his twilight haunt. The plaintive cry of the veery from the tree-tops above only serves to emphasize its silence, while the scream of its warder, the blue jay, seems its voice speaking to the solitude. I usually find what might be termed a foot-path threading a swamp, not always readily discernible, but sufficiently marked to make it appear a foot-path, the highway of the hares and wild animals. These resort to it not only for food and water, but for warmth and security. The hibernating birds turn to it instinctively and seek it for their winter quarters.

The swamp is Nature's sanctuary—the great gamekeeper and game-protector. It is the rampart of the landscape. Within its sheltering arms is nurtured the most beautiful of sylvan utterances, the roll-call of the ruffed grouse. Without its helping hand both furred and feathered game must in many localities become virtually exterminated, and a wood without game is a wood devoid of one of its most individual attributes. There is ever a charm in the elusive, the untamed in nature; to have its wild animate forms about us, though we may only clasp the shadow. The trout-stream in its mazes through the woods possesses an additional voice and meaning to me for the radiant life that lurks within its pools and shallows. I care less for the rod than to feel the rightful habitant is at home.

The owl's weird cry borne upon the December dusk without brings the wintry woods into my room—the rustle of dry beech-leaves, the breath of lichens and of pines. All Nature for the instant seems articulate in his cry. You may never meet the fox face to face unaided by the hounds; but it is a satisfaction to know he is present. Keen of scent and fleet of foot he has passed long before you, evaded you; yet he is there, somewhere, farther on amid the mystery and silence, in all his lissome grace and suppleness of sinew. The very footprints of the hare recall the living presence of the hare, his wild beauty and his nimble speed. So that in a swamp or wood tenanted by game this fascination is ever present—the living unconfined creatures appearing a component part of the trees and undergrowth, with which they blend and become incorporated, just as the shadows belong to and accentuate the strength of the sun. So also in the garden copse, when the mold is starred with *Hepaticas* and *Trilliums*, the wild flowers are obliterated for the moment to me when a squirrel barks or a white-throated sparrow sings.

In the swamp, on blustering days without, I

see the downy woodpecker's scarlet coronet, his busy mallet beating its sonorous rat-tat-tat on hollow trees. I catch, too, the fine call-note of the little brown-creeper running up and down and around the limbs and tree-trunks in quest of his food, and hear the flute-like call of the tree-sparrow feeding on the spicy buds of the sweet birch. I mark the caressing "day, day, day" of the black-cap chickadees, happy in the cold and storm, while the solemn "yank, yank, yank" of the nut-hatch is never still. Leaving the woods proper on a windy winter's day, even a sheltered beech-wood where the clinging foliage of the beeches and hornbeams wards off the wind, there is an ever-fresh surprise in the absolute absence of wind and positive warmth of the swamp. Green as in midsummer are its club-mosses and evergreen ferns, and the goldthread, wintergreen, and partridge-vine seem merely hibernating beneath the snow. A temperature it possesses of its own—cool in summer and warm in winter—and a flower I find cradled in its shade always appears to have gained in purity or refinement of hue.

Another shade-loving plant now passing out of blossom is the white swamp honeysuckle (*Azalea viscosa*), succeeding the pink *A. nudiflora*, whose fragrant flower-clusters, exhaling the characteristic honeysuckle odor, proclaim its presence. The tall red lilies along the edge of the swamp have long since made their summer display; but the fading flower-spikes of the greater orchid are still seen in low places just as the ladies-tresses are forming their flower-heads amid the meadow grasses. The spring beauty and *Trillium* have vanished from the woods, and *Hepaticas* and *Violas* are hidden by the stronger growing plants of midsummer. There is a crowd of tall evening primroses, white and purple *Eupatoriums*, pink *Epilobiums*, blue vervains, pale asters, yellow golden-rods, and helianthuses, all jostling and striving for supremacy.

Growth is rank on every side. It is the seed-time and harvest of the big weeds, when the waste places become a veritable jungle, perilous and almost impassable to man and beast. It is the high carnival of sticktights, nettles, burdocks, briers, brambles, tares, thistles, teasels, and *noli me tangeres* innumerable, among which the true touch-me-not or jewel-weed least deserves its name, for there is nothing noxious about it or vicious in the strange bursting of its seed-pods at the touch, whence it derives its appellation. The sticktight, the tare, and the burdock are the true fiends incarnate among the sticking and stinging weeds. I revere the inventor of corduroy, the only coat of mail with which one can wade comparatively unscathed through the gantlet of these tramps and ruffians of the field.

The everlasting is white with flower in the pastures, and on sunny upland slopes rank upon rank of mullein-spires tower above the carpet of fragrant pennyroyal. Along the water-courses *Heliopsis lævis* has set its fringe of gold, visible from afar, the *avant-courier* of the pageant of autumn that will come in a tidal wave of color to brighten the declining year.

The Hardy Fernery.

You will pardon some obscurities, for there are more secrets in my trade than in most men's. And yet not voluntarily kept, but inseparable from its very nature. I would gladly tell all that I know about it, and never print "No admittance" on my gate.—THOREAU.

XII.

THE HARDY FERNERY.

WHATEVER the garden may owe to hardy flowers, and however varied and attractive its collection of shrubs and trees, it would still be lacking in one of its greatest charms if deprived of ferns. They are the very quintessence of the woods, whether they rise to form a classic urn like the great ostrich, or quiver on ebon stems like the lovely maidenhair. The very name has a fresh, fragile sound in any language — *Filices, felci, fougères, Farnen, ferns.* The fern offers no excuse for not possessing flowers. Color, other than its infinitely varied greens and the dark spore-cases underneath or on the margins of the fronds, would mar its beauty. Its green and its grace are its flower, and Nature wisely left it a flowerless plant, the embodiment of beauty in foliage. When well grown the fern carries its characteristic tropical effect to the garden, and, once established, the hardy fernery may become one of the finest ornaments about the home. It is, however, seldom seen to good advantage under cultivation, for the simple reason that it is generally left to take care of itself, a matter it is never called upon to do in its native state, where it is protected from wind, has its fronds moistened by condensation, and is provided with congenial soil and coveted shade. The delicate beauty of a fern-frond can not be obtained outside of its native habitat without in part reproducing the natural conditions under which it grows. Shade, shelter, moisture, and suitable soil are its main requirements. Some species, of course, occur naturally in sunshine and dry soil, and these may be grown under like conditions. The long period during which they retain the freshness of their fronds is a notable feature of the genus, while, whatever the season of the year, some of the species are found perennially green. Most hardy ferns are not difficult to cultivate, many being very accommodating and growing where little or nothing else would. Hot summers do not affect them disastrously as is the case with many plants, providing sufficient water be supplied at such times.

On the north side of the house, beneath the

shade of non-surface-rooting trees and in low, moist positions, a very large number of hardy native species may be successfully grown. Not a few of the species, even those which naturally occur in shade, will do well in open places, though, except some of the sun-loving kinds, few will attain that luxuriance and delicacy of color they possess in shade or partial shade.

A shady and sheltered position will, therefore, be chosen for the hardy fernery; for shelter from winds is no less important than protection from the direct rays of the sun. This position should be readily accessible to a fine dust-spray attached to the hose. Ferns are generally found in moist situations, thriving in a humid atmosphere; and these conditions must be followed as nearly as possible. But while ferns and moisture are almost synonymous, constant watering is, nevertheless, to be avoided. It is only when the soil is becoming dry, before the dryness is felt and shown by the sensitive fronds, that watering is necessary. The foliage of ferns does not like constant drenchings, pelting rains frequently being as injurious as severe winds. But the effects of wind are more severely felt where the plants do not receive their necessary supply of moisture, the stems becoming more brittle if the roots are not moist and cool. Watering the grass and the surroundings of the fernery in the evening, when the ferns themselves do not require watering, is appreciated by the plants, this tending to preserve a humid atmosphere. Watering a little every day or two merely keeps the surface damp, and does not reach the roots, or prevent the foliage from becoming dry. It is far better to give a good supply of water occasionally, as the plants require it; an observation that will apply equally to most other hardy plants. Having chosen the position for the fernery, the ground should be dug to the depth of two feet, and filled in for the most part with black muck, leaf-mold, and a small portion of sandy loam. This gives a light, elastic soil, retentive of moisture and suitable for most ferns. The fernery is much benefited by a liberal top-dressing of old leaf-mold every autumn; and, aside from the protection to some of the less hardy species, a thick winter covering of leaves and evergreen boughs is advisable, in order to prevent the heaving of the ground by frost.

The common ostrich-fern (*Onoclea struthiopteris*) is among the most robust and easily grown of the genus, which numbers in the United States some one hundred and sixty-one species, fifty of which are indigenous to the State of New York. On account of its strong growth and the frequency with which it throws out suckers from its rambling rhizomes, it is best placed by itself. Planted numerously with other species it soon crowds them, unless the suckers are checked. Few plants have a more tropical effect than this, a mass of it forming a grand feature of any garden. It is well and tersely described by Gray— "a fern of noble port." This does best in shade, but it may also be grown in sun.

The royal fern (*Osmunda regalis*), which belongs to the class of flowering ferns, is rarely seen to good advantage under cultivation. It is, likewise, one of the most robust of the genus, occurring naturally both in open sun and dense shade, but always in wet or moist situations. Perhaps there are none of the large species whose color varies so much, the young plants, more especially in sunny situations, assuming varied shades of reddish-green. In rich, marshy places it frequently grows to a height of five feet. It is pre-eminently a bog-garden plant, where it may be grown as vigorous as it occurs naturally, the bronze and copper hues showing more boldly in open situations. A smaller form (*O. gracilis*) occurs, with broader foliage and more urn-shaped than the type.

A very common fern, found in dry places, is another of the same species, the interrupted flowering fern (*O. Claytoniana*), interrupted near the center of the leaf-stalks by several pairs of fertile leaflets densely covered with brownish sporangia. This gives a rusty, unfinished look to the fronds, and for this reason it is undesirable for the fernery, and not to be compared with another of its family, the cinnamon-fern (*O. cinnamomea*). The yellowish fertile fronds of this, springing from the center of the plant, during its younger stage, are distinct and beautiful, while the species is a tall, robust grower.

Perhaps the most distinct of native ferns is the sensitive fern (*Onoclea sensibilis*), common to low woods and moist grounds. Aside from its striking peculiarity of foliage and its dark-colored spore-cases, its young fronds, throughout the summer, wear a lovely light-green hue possessed by no other member of the genus. The sensitive fern should be grown in shade, the fronds quickly becoming scorched by sun. It would impart a distinct appearance to the garden landscape grown *en masse*, being so rarely seen in gardens. It is one of the best ferns amid which to plant the tall wild red lilies. Owing to its being somewhat tardy to start into growth, the latter do not become choked, as they are by the more forward and ranker-growing ostrich.

The common brake or bracken (*Pteris aquilina*), while distinct from the generality of ferns, is not worth cultivating, unless on the margins of woods, or places where little else will thrive. It spreads with great rapidity, and soon becomes a pest if placed among other ferns. The big moonwort (*Botrychium virginicum*), the largest of the species, differs essentially from most of the genus. It is termed "a beautiful fern," but does not show to advantage when planted with others of its tribe.

The shield-ferns, or *Aspidieæ*, number many of the noblest of hardy ferns. Of these, the deciduous *A. aculeatum* and *A. Goldianum*, the evergreen *A. achrosticoides*, *A. cristatum*, *A. filix-mas*, *A. marginale*, and *A. spinulosum* are among the finest, best known, and most easily grown. Nearly all of the species, whatever their size, are delicately beautiful, the finely serrated plumes being a conspicuous characteristic. The woods where I find the ruffed grouse and the large white hares in winter would seem lonely without the freshness of the Christmas-fern and the perennial verdure of the evergreen wood-fern. The frost, whose sharp scythe has cut off the foliage and the flora, seems only to have brought out a richer green in these flowerless plants, that never look half so lovely as they do in winter. They seem the type of hardiness and longevity, and mask the loneliness of the leafless trees.

Every one knows and admires the maidenhair (*Adiantum pedatum*), its fragile, polished stem supporting its delicate lace-work of foliage. Erroneously supposed to be difficult to cultivate, the maiden-hair, nevertheless, takes quite kindly to cultivation when placed amid congenial surroundings and allowed time to become established. Two among medium-sized ferns— *Cystopteris fragilis* and *C. bulbifera*—deserve a place on the front edge of the fernery. If the former has a fault, it is the early discoloration of the fine fronds. But it is one of the most graceful of its tribe, as well as one of the most forward to clothe with green the bases of trees in the woods of early spring. *C. bulbifera* is less common, but very prolific where it occurs—a delicate fern, with long, slender, arched fronds. I have found this troublesome in the rock-garden, on account of its coming up almost everywhere soon after being introduced.

There are numerous other desirable species, of large and medium habit, that may appropriately find a place in the hardy fernery; but, for all ornamental purposes, a sufficient variety may be obtained by those already specified, without further extending the list. It is, perhaps, superfluous to remark that where the fernery is placed by the side of the house, or against a wall, the more robust kinds should occupy the background, and the smaller-growing species the foreground, where they can not become smothered. Thus far I have referred only to the more robust species. But a great merit of the *Filices* is, that the smaller they become the more beautiful they seem. The little oak-fern (*Phegopteris dryopteris*), for instance, whose delicate print is found on decayed logs and moist, shady places, is one of the loveliest of its family. The diminutive polypody, too, that drapes dry bowlders with its living green, is a fern one must always stop to admire, however common it may be.

These smaller ferns, with many others, can not be grown with the larger sorts, and must have a special place, either the rock-garden proper or a small bed by themselves. The oakfern and beech-fern are easily established in leaf-mold and loam. The common polypody and the larger and handsome *Polypodium falcatum* are not always so accommodating, preferring a mixture of peat, leaf-mold, and sharp sand or sandy loam. There are very many varieties of the polypody cultivated in England.

Woodsia Ilvensis and *W. obtusa* are beautiful small ferns.

The curious walking-fern (*Camptosorus rhizophyllus*) I have found difficult to establish, and the charming little maiden-hair spleenwort (*Asplenium trichomanes*), though numerous specimens of it live on from year to year, never looks quite vigorous. *A. ebeneum*, a larger species from Oregon, I have found rather fastidious, as also *Cheilanthes vestita* and the delicate *Cryptogramme acrostichoides*. The distinct hart's-tongue (*Scolopendrium vulgare*) does well with me. Upward of fifty forms of the latter are cultivated in England, many being of marked beauty. *Asplenium nigrum* is an easily grown small English fern which will grace any collection. *Celerach officinarum* is likewise a very distinct and handsome small British fern, though not so easily grown as the latter.

To grow the more delicate small ferns successfully demands a favorable climate and location with a thorough knowledge of their requirements, and only true fern-lovers who are willing to devote the necessary time and study will find it worth while to attempt the cultivation of the greater portion of the very beautiful smaller *Filices*.

It is more satisfactory to collect ferns yourself; they then become a pleasing reminder of many a locality where they were obtained. Removal may be successfully effected at almost any season. For beginners early autumn is a favorable time for collecting, as it is near the dormant season; and yet the various species may be readily distinguished, the fronds having not yet dried.

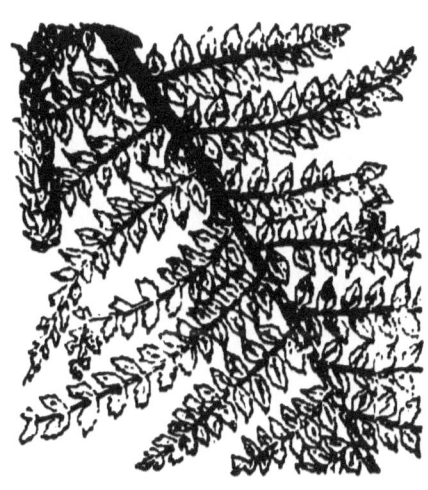

Midsummer Flowers and Midsummer Voices.

That time of year, you know, when the summer, beginning
  to sadden,
Full-mooned and silver-misted glides from the heart of September,
Mourned by disconsolate crickets, and iterant grasshoppers
  crying
All the still nights long, from the ripened abundance of
  gardens.
HOWELLS.

XIII.

MIDSUMMER FLOWERS AND MIDSUMMER VOICES.

AFTER blossoming profusely throughout latter June and the first half of July, the Japanese honeysuckle, as if to emphasize its attractions, again bursts into delicious bloom during late August and September. The tiger-lilies have been constant through late July until late August, when most of the species have passed. But the crowning glory of the lilies is *auratum*, which extends its blossoming period throughout August and September, no species of the genus being so continuous to blossom. The odors of *Lonicera Halleana* and *Lilium auratum* are not unlike, and numerously planted in front of the verandas they flood the whole house with perfume in the evening. A beautiful flower becomes doubly beautiful when it prolongs the usual flowering season, and, judged by this standard, both the Japanese honeysuckle and golden-banded lily deserve our warmest praise.

Some of the roses also are flowering for the second time. Among them I mark especially Marguérite de St. Amande, Marshall P. Wilder, and Paul Neyron. With them the following may be named as among the most free-blooming autumnal sorts: Comtesse de Serenye, Rev. J. B. M. Camm, Boieldieu, François Michelon, Mabel Morrison, Louis Van Houtte, La Reine, John Hopper, Baroness Rothschild, Baron Prévost, Countess of Oxford, Eugénie Verdier, Marie Beaumann, Victor Verdier, Hippolyte Jamain, Horace Vernet.

Very companionable during August and September are the althæas, almost the only flowering shrubs blossoming at this time. In the rear garden there is a swarm of bright flowering annuals—petunias, verbenas, calendulas, escholtzias, nasturtiums, and marigolds. Herrick tells how marigolds came yellow:

> Jealous girls these sometimes were
> While they lived or lasted here:
> Turned to flowers, still they be
> Yellow markt for jealousie.

This may apply to the orange-colored kinds, not to the big double lemon-yellows, too handsome to be jealous of any flowers of their color.

For weeks there has been a notable absence of bird-voices. The English sparrows are for the most part on a vacation to the grain-fields. The songsters are almost silent save the constant wood-pewee, who, however, only utters the first two notes of his plaintive cry. His is a haunting, melodious strain I should sadly miss from the copse and garden. The ornithologists describe his voice very variously. Coues speaks of the "sobbing of the little somber-colored bird"; Wilson places him "amid the gloom of the woods, calling out in a feeble, plaintive voice 'peto-way, peto-way, peto-way'"; Langille terms his notes "a slow, tender, and somewhat melancholy whistle, 'pe-wee'"; Flagg refers to his "feeble and plaintive note"; Trowbridge, in his poem, interprets his song, "Pe-wee! pe-wee! peer!" Burroughs alone rightly describes it as "a sweet, pathetic cry." It is, in addition, a cry of considerable volume and penetration, its sweetness masking its real force, always plaintive, and, when the full strain is delivered, wonderfully effective at the close. I can not discern anything resembling "pe-wee" in either call or response unless it be in late summer. It sounds distinctly *whē-ū whē; whēē ū.*

The common pe-wee or phœbe-bird possesses no such subtle charm. He never tires of reiterating the two notes of his refrain. It sometimes tires the listener, however, and a misogynist might wonder if it is not the female who sings.

To compensate for the silence of the birds, the insect world is shrilling *con amore* night and day. So many instruments compose the orchestra that one is puzzled to place all the performers. Loudest of all is the cicada's great *crescendo*, overpowering the strumming of grasshoppers and droning of diurnal crickets. The shrill of the common black cricket, produced by rubbing his legs sharply together, consists of three notes in rhythm, and is said to form always a triplet in the key of B. Night is the morning of the green leaf-cricket's day. At twilight or late afternoon he begins his even-song in strong, well-modulated notes, chanting continuously until daylight. His chorus it is we hear so steadily, commencing briskly in August, and uttered, now fast, now slowly, according to the warmth or coolness of the night. His voice is extremely deceptive, appearing to proceed from almost any place except the vine or tree overhead. A plaintive, soothing song he sings, a song in keeping with the season, pulsating with every change from heat to cold, and finally subsiding to a scarcely audible sob in late October.

In the fields *Grylliadæ* innumerable are loud with song. Listening to the melody of their countless wings, strange it seems that their transitory existence is but the enactment in another world of the passions and jealousies of our own; that their *allegro* of stridulous sound is but an expression of the fierce rivalry of males; that the grasshopper's voice proceeds from a stamping-ground of strife, and the "crink-crink" of crickets is largely the declaration of jealousy and hate.

From the raspberry-vines rises a dreamy, summery voice, continuous during the day and not unfrequent during the night, proceeding from one of the small climbing crickets. Up go the long antennæ and gauzy wings, and a prolonged

"Crēē-ē-ē-ē-ē-ē-ē-ē-ē-ē-ē-ē-ē-ē-ē-ē-ē"

trembles upon the air. This is *Œcanthus fasciatus*, one of the pleasantest of insect-choristers. How his delicate wings withstand the constant scraping they do, and how they can produce such a clear, bell-like sound, seems inconceivable. Like the green leaf-cricket, he is an accomplished ventriloquist. One of these, having escaped from confinement, and singing unceasingly, led me a twenty minutes' search ere I could locate his precise whereabouts from the song, which seemed everywhere but the exact spot whence it proceeded. There is another insect, not a troubadour, who adds his harsh note to the orchestration of the hot nights of midsummer and early fall. His stridulation possesses the characteristic rasping of the katydid tribe, but is less intense. On cool nights he is silent, but as soon as the nights become warm he commences to file his saw until dawn.

Of all familiar insect sounds, the voice of the cicada is the strongest, that of *Œcanthus fasciatus* the most summery, the green leaf-cricket's the most plaintive, and the katydid's the harshest. The general effect of all these minstrels, save that of the katydid, is a soothing one. The bird-songs of spring are happy, merry, buoyant, I may say, wakeful—a triumphant major of song. The insect-chorus of fall is an ode rather than a lyric—a song pitched in a minor key, rising and falling amid the lengthening shadows and gathering haze of autumn.

Latter midsummer and early fall bring a fresh color-surprise to the garden. It is the season of the phloxes, tritomas, helianthuses, the great hydrangeas, the Japanese anemones, and the stately autumnal flowers—the gathering and concentration of months of warmth and sunshine. One expects much of the late flora, it has been so long about its task. Less of grace and tenderness it possesses than that of spring, but greater strength of stalk, and more of boldness and virility. The phlox, a genus exclusively North American, constitutes, in the large garden species, one of our most varied and valuable hardy perennials. America has furnished the phloxes, Europe has hybridized them; the garden perennial phlox, as it is now perfected, having originated from the tall-growing *P. paniculata* and its varieties, and the lower-growing *P. maculata*. That such brilliant varieties as "Lothair," and some others, are the result of a cross with *P. Drummondii*, would not seem improbable. The *decussata* class embraces the taller-growing varieties, the *suffruticosa* the smaller.

The hybridized phlox has its own gamut of colors, like the hybridized azalea—hues and tints possessed by no other flower. These glowing shades of salmon, rose, and vermilion, together with the numerous pure white and creamy-white varieties, are the more striking from the grand flower-trusses and the tall stalks upon which they are placed. The phlox may be termed a necessary garden-flower. It is easy to grow, of hardy, vigorous habit, and there is no other perennial to supply its place. The largest trusses are produced on two- and three-year-old plants. Renewal in some soils becomes necessary every few years. Where it thrives, however, the magnificent head of bloom carried by an old plant is far more showy than the few larger trusses of a younger one. Pinching or cutting back the shoots in early June will render it autumnal flowering, and by thus treating a portion of the plants the flowering season may be largely prolonged.

There is another advantage from pinching some of the plants : if the weather be unusually hot and dry during August, so as to cause poor flowering, the retarded plants will almost always have the advantage of cooler weather to flower. Pinching, however, is done at the expense of the size of the truss ; where large trusses are an especial object, at least half of the stalks should be cut out. The phlox needs abundant moisture during its florescence, and likes rich soil. Some phloxes, particularly the whites, are subject to mildew in certain soils and during certain seasons. But the great enemy of the phlox is the larval grub of the big May beetle, with whom the roots of this plant are an especial favorite. His presence may be detected by the sickly appearance of the plant—a knowledge that usually comes too late. Young plants should be immediately lifted, the grubs destroyed, and the plants replanted in fresh earth. Old phloxes being impatient of removal, there is seldom any remedy when they are thus attacked. There are very many beautiful varieties of the tall-growing perennial phlox, varying more or less in strength of stalk, size of trusses, and length of time the plant continues in vigor. Among the desirable kinds which are due to the skill of the hybridizer, it may be well to specify some of the most vigorous and distinct, the extended catalogue lists being somewhat confusing to the amateur. For the phlox, like numerous other florists' flowers, is not without crying colors, its purples and rose-purples in particular being generally objectionable. Of the whites, which are indispensable to set off the hues of their companions, queen of the whites and Viérge Marie are the best, the latter being a taller grower, later to blossom, and better withstanding hot sun. Of the light-colored varieties Richard Wallace, white with violet center, and Premier Ministre, white with rose center, are among the strongest growers, and produce the finest trusses. Of the lighter reds perhaps the most distinct are Gambetta, Madame Lechurval, Rêve d'Or, L'Avenir, and Lothair, the latter the most brilliant of the salmon-colored forms, and one of the most brilliant of flowers. Parrai, recently sent out, is the finest of the rose-colored type, while Oberon still remains the finest deep red.

The great Chinese plume-poppy (*Bocconia cordata*) is a very handsome late-flowering plant. It attains a height of nine feet, and the large terminal flower-panicles and tropical oval-cordate leaves are extremely graceful and showy. But it is a rambler at the root, and must have a place where the suckers will not cause trouble. It is not a safe plant for the border or the lawn on this account, where otherwise it would be highly ornamental.

The Japanese *Polygonum cuspidatum* is another grand herbaceous plant that is tempting to employ, but which he who is wise in his generation will avoid. Once established, it becomes a horrible nightmare, and I doubt if there exists among hardy plants a more troublesome subject to banish. My experience has been confined to a colony in my neighbor's garden, close to the division-line, that from year to year extended its deep-rooting suckers farther and farther on to the lawn and borders. I shudder now when I think of the digging and trenching and undermining and the barrels of salt it has required to prevent its intrusions.

I mentioned this pest to a friend noted for his marvelous knowledge of hardy plants, and for his splendid garden at Edge Hall, Cheshire. For once he was caught napping, and opened his garden-gate to a wolf in sheep's clothing. "I was younger than I am now," he remarked, with a smile illuminating his splendid face, "and have had fifteen years' experience with *Polygonum cuspidatum*. It established itself in one part of my garden so that it has kept me busy fighting it for years; and a man still works half a day every fortnight in the vain attempt to eradicate it."

A fine, old-fashioned flower is the white day-lily (*Funkia grandiflora*), with pure white, fragrant blossoms during August and September—a flower almost too common and well appreciated to need specification. The curled-leaved variety of the common tansy (*Tanacetum vulgare, var. crispum*) is well worthy a place in the flower or shrubbery border. Its scented leaves are refreshing to smell as you pass, and are as beautifully curled as the fronds of the crisped hart's-tongue fern. I had almost overlooked the garden thyme, now forming great cushions on the rock-work. It is aptly named from *thumos*—courage, strength—the smell of thyme being reviving. The variegated-leaved varieties—the gold- and silver-leaved—are the most desirable, always elegant in the rock-garden or flower-border. A singular flower at this season is that of one of the tobacco-plants (*Nicotiana affinis*), which opens its long, pale trumpet in the evening, exhaling a rich odor like that of the petunia. This scent it withholds during the day. When cut, the flowers remain open in the house, scentless during the day, and becoming odorous at night. The plant itself, an annual, can scarcely be considered ornamental, and its leaves are a great favorite of the potato-bug. But at dusk the numerous long-tubed white flowers are very conspicuous and striking, and it well merits a place in the rear garden for cutting purposes. *Nicotiana tabacum*, the cultivated tobacco, and its varieties, are fine tropical-looking plants, with huge leaves, that may be employed to excellent advantage as foliage-plants, in company with the eulalias, and the taller growing grasses and rushes. But, being annuals and tender, they require to be raised from seed sown early in the spring. The evening primrose is likewise a curious flower, lighting its pale-yellow lamps in the evening and extinguishing them by day. The flowers open audibly, the expanding of the corolla being characterized by a peculiar sound as the flowers pop open. Besides the common evening primrose (*Œnothera biennis*), a troublesome weed in many States, there are numerous other species to which garden space may well be accorded. The larger flowered variety of *biennis*, called *grandiflora*, is to be preferred to the type, *Œ. fruticosa*, and its varieties grow from one and a half to three feet high, and are abundant bloomers with smaller, bright-yellow flowers. *Œ. speciosa*, a fine species of Arkansas and Texas, and a dwarfer form, bears numerous large white blossoms passing to rose. *Œ. Missouriensis*, a Western species remarkable for the size of its flowers and fruit, is a small-growing form well adapted for rock-work, with vivid golden-yellow blooms appearing from July to October. The evening primrose is far better known in English gardens than it is at home. Its color, fragrance, and free-flowering habit render it a desirable plant of late summer, when it is grown in rich soil. Only some of the species are nocturnal, though most of them are more odorous, and open more fully in the evening. All the tall-growing kinds seed freely, and are readily grown from seed.

August and September are the months of the sunflowers, or *Helianthea*, named from *helios*, the sun, and *anthos*, a flower, from the erroneous but common opinion that the flowers always turn their faces toward the sun. The appellation is appropriate, notwithstanding; for there are few brighter, more sun-loving flowers than this extensive tribe of the composites. The species grow mostly from four to twelve feet high, and are characterized by their large, showy, yellow flowers, the largest being *H. annuus*, the well-known Peruvian annual. The *Helianthus* is coarser than numerous other garden favorites; and while many of the species undoubtedly are better adapted for the wild garden, there are still a number well deserving a place in the flower-border and shrubbery. To combine size, hardiness, and luxuriant bloom, one must sometimes put up with coarseness; and any weedy appearance of the perennial sunflowers is more than atoned for by the gayety many of the species impart to the garden at a time when they are really required. To the sub-tribe of the *Helianthea*, included in the tribe *Asteroidea*, belong also the *Heliopsis, Rudbeckia, Echinacea*, and *Coreopsis*, mostly perennials in the style of *Helianthus*. From all of these we have a great mass of yellow autumnal blossom not to be dispensed with. All the sunflowers grow well in any common garden soil, most of them being easily raised from seed, while many readily form hybrids. Even the big annual deserves a space somewhere in the rear garden, and merits the encomium of Clare:

> Sunflowers, planted for their gilded show,
> That scale the lattice windows ere they blow.

In the mythology of the ancient Peruvians it occupied an important place, and was employed as a mystic decoration in ancient Mexican sculpture. Like the lotus of the East, it is equally a sacred and an artistic emblem, figuring in the symbolism of Mexico and Peru, where the Spaniards found it rearing its aspiring stalk in the fields, and serving in the temples as a sign and a decoration, the sun-god's officiating handmaidens wearing upon their breasts representations of the sacred flower in beaten gold. Numerous varieties of the great-disked sunflower exist. It is the art of the gardener to know how to place them. I turn to old Gérarde to find him an enthusiast over the great flower of gold. His cultural directions still hold good: "The seed must be set or sowne in the beginning of Aprill, if the weather be temperate, in the most fertile ground that may be, and where the sun hath most power the whole day." And where is the florist or botanist who can paint the marigold of Peru as vividly as it was etched in the "Great Herball," or "The History of Plants," nearly three centuries ago? Almost all of Gérarde's, as also most of Culpepper's icons of plants, are models of their kind, bringing the plant before one, not only truthfully, but picturesquely. You see its form and color, savor its fragrance, become acquainted with its virtues— you fairly see the plant grow and the flowers expand. The page was ampler in the days of the colon and the tall folio, the margin for embroidery wider, and author and reader were less hurried. To-day, after the lapse of centuries, the descriptions stand out with the vividness of an old copper-plate proof. The reader who has had the patience to follow me, and who does not know him, will be interested in a typical description by Gérarde: "The Indian Sun or the golden floure of Peru is a plant of such stature and talnesse that in one Sommer being sowne of a seede in Aprill, it hath risen up to the height of fourteen foot in my garden, where one floure was in weight three pound and two ounces, and crosse overthwart the floure by measure sixteen inches broad. The stalkes are upright and straight, of the bignesse of a strong mans arme, beset with large leaves even to the top, like unto the great Clot Bur: at the top of the stalke cometh forth for the most part one floure, yet many times there spring out sucking buds, which come to no perfection: this great floure is in shape like to the Cammomil floure, beset round about with a pale or border of goodly yellow leaves in shape like the leaves of the floures of white Lillies: the middle part whereof is made as it were of unshorn velvet or some curious cloth wrought with the needle; which brave worke if you do thorowly view and marke well, it seemeth to be an innumerable sort of small floures, resembling the nose or nozell of a candlesticke, broken from the foot thereof: from which small nozell sweateth forth excellent fine and cleere Turpentine, in sight, substance, savour, and taste. The whole plant in like manner being broken, smelleth of Turpentine: when the plant groweth to maturitie, the floures fal away, in place whereof appeareth the seed, blacke, and large, much like the seed of Gourds, set as though a cunning workeman had of purpose placed them in very good order, much like the honiecombes of Bees: the root is white, compact of many strings, which perish at the first approach of winter." What more could be said of the plant he is depicting, unless by the bees who draw nearer to the heart of the flower than we? And who could depict it half so well! Plant-knowledge is assuredly more accurate since the Linnæan and natural systems, but plant-study isn't half so picturesque as it was when the old masters held the magnifying-glass. And, after all, who will object to an error when the picture is so artistically painted? Is not a misnumbered page a charm of an Elzevir? There are some much smaller ornamental annuals than the large Peruvian species, but so many superior forms of perennials exist, that the former are scarcely worth the trouble of growing.

Among the earliest of the perennial sunflowers is the showy ox-eye (*Heliopsis lævis*), frequent along streams and banks, where its numerous yellow flowers form vast golden streamers during August, conspicuous from a great distance. At nearly the same time *Helianthus divaricatus* peoples the thickets and meadows—a brilliant lemon-yellow flower. Later comes *H. decapetalus*, the blossom not unlike the preceding, but the plant more bushy and more numerous flowered. *H. multiflorus* bears showy yellow heads, there being a major variety of this superior to the type. The double form (*H. m. flore-plenus*), much seen in gardens, is among the most conspicuous of perennials, carrying a huge sheaf of golden bloom, the large double flowers remaining long in perfection. *H. multiflorus* increases very fast, a small root set out in early spring forming a large bush by August. The flowers are always larger on young plants; after the second year they diminish in size, when the plants should be divided. *Rudbeckia hirta*, the orange-daisy of the fields, would be desirable were it not so common a weed. *R. speciosa* or *Newmani* is a preferable form, being a stronger plant, and less straggling in habit. *R. nitida*, a taller-growing plant, is one of the best of its class, extremely showy in masses, though it lacks the dark contrasting cone that characterizes the other forms specified.

Much resembling some of the perennial sunflowers is *Silphium perfoliatum*, one of the several coarse, tall-growing rosin-plants, flowering in July and August. It has huge leaves, great clusters of large lemon-yellow blossoms, and grows seven to ten feet high. For stately and brilliant effect it has no rival during latter August, forming a golden cloud of flower, the more striking from its tall stalks and deep-green foliage. The most remarkable of the genus is *S. laciniatum*, the compass-plant of the prairies, which is said to have the peculiarity of turning the edges of its lower leaves north and south, but this is not noticed in cultivation. This grows from eight to twelve feet high, having large yellow flowers and immense leaves. Other species are *astericus, integrifolium, trifoliatum, terebinthinaceum,* and *asperrimum,* all with yellow flowers, and *albiflorum,* with white flowers. The proper place for most of the *Silphiums* is the rear garden, or the edge of a distant shrubbery, in masses. But *S. perfoliatum* is always worthy of a topmost seat in the garden synagogue, when in bloom. The *Heleniums* are tall-growing plants, with large yellow or orange flowers, similar to *Rudbeckia. H. autumnale,* the most common of the genus, is a conspicuous plant, growing from two to three feet high. *H. Hoopesi* is a coarse plant, growing three to four feet high, flowering in August and September. The flowers are showy, produced in umbels; the color of rays and disk is uniform bright orange. *H. pumilum* is the smallest and perhaps the poorest of the genus, none of which take the place of the *Helianthus* or some of the *Silphiums.*

Helianthus rigidus, generally known as *Harpalium rigidum,* is a very striking early species, with large, brilliant, dark-yellow flowers. It is the lowest-growing of the genus, not exceeding four feet in height. Its habit is to run much at the root, and therefore it soon becomes troublesome in the flower-border. But it should not be neglected on this account, and a place should be found for it where it will have room to make its bright midsummer display. The *Echinacea*, or cone-flower, on account of the large heads of purple produced by *E. purpurea* and *E. angustifolia*, is worth growing. *Coreopsis lanceolata* is the finest of its genus, although *C. grandiflora*, *C. pubescens*, and *C. auriculata* are not unlike it. *C. verticillata* is a small and pretty species, with delicate foliage and numerous small yellow flowers. *C. præcox* is scarcely worth growing, notwithstanding it is described as "*cette charmante espèce*" in the suave French catalogues. Other fine *Helianthea* are: *Helianthus doronicoides*, *H. strumosus*, *H. orgyalis*, *H. giganteus*, *H. tuberosus*, and *H. Maximiliani*; the last three growing from nine to twelve feet high. *H. giganteus* has purplish stems, rough, hairy, lanceolate, and sessile leaves; flowers two and a half inches across, abundantly produced in August. *H. doronicoides* is one of the finest of the sunflowers, a large-flowered, large-leaved, tall-growing species, with bright-yellow blossoms. *H. Maximiliani* requires a warm climate to show flower, it being the latest of the genus to blossom.

At this season spiders become very annoying in the garden, weaving their webs among the flowers and leaves, so as to give an untidy appearance to the shrubs, vines, and flower-borders. They may serve some subtle purpose besides catching flies, these hordes of weavers, big and little, white and brown. But their dust and leaf and insect and pollen-strewn shuttles are certainly unclean. The sparrow will not walk into their parlor, and brushing away the webs or drenching them with the hose is merely temporary. There is but one way to treat the September spider—to follow the example of the mistress of the house, and kill him, cruel as it may appear.

Flowers and Fruits of Autumn.

LET the shadow advance upon the dial—I can watch it with equanimity while it is there to be watched. It is only when the shadow is not there, when the clouds of winter cover it, that the dial is terrible.

    RICHARD JEFFERIES, THE LIFE OF THE FIELDS.

XIV.

FLOWERS AND FRUITS OF AUTUMN.

PLEASANT weather glides by so swiftly in the garden, that September is well advanced toward October before we realize that the year has begun to wane, and the flowers have but a brief period to stay. Yet a glance at the border reveals no such apparent consciousness on their part. June, with all her exuberance of bloom, scarcely contributes a grander floral display than do the stately flowers of September—the *Helianthea*, the decussata phloxes, the perennial asters, the Japanese anemones. The lavish odors and delicate hues of the early season are lacking in the later flowers; a ripe brunette in yellow has come to take the place of the fair young bride of Spring. A full supply of moisture during the first part of the month has had a beneficial effect in prolonging the freshness of vegetation. It is only the ever-increasing chorus of crickets, the lengthening shadows, and an admonitory rustling in the fast-ripening leaves of the trees that point inexorably to the hour of the year.

For the second time many of the larkspurs are blossoming. In the rock-garden the little star-grass, *Viola pedata bicolor*, and the white and purple *V. cornuta* simulate another spring. *Campanula Carpatica* is covered with bloom; *Tunica saxifraga* has not ceased to blossom since early summer, and *Spiranthes cernua* remains the sweetest wild flower of September. The *Cannas* and the great Japanese variegated grasses are just attaining their full beauty. When planting *Eulalias*, the smaller-growing *E. gracillina univitata* should not be omitted, an extremely distinct and beautiful species with wiry, grass-like foliage.

But the garden will not take care of itself even now; to preserve its fresh appearance the knife must be frequently employed to remove the withered stalks, and the rake to collect the fallen leaves. Left to themselves, the borders would already look rusty in August, and removing withered leaves and stems forms no small portion of the season's work. The stems of many perennials should not be cut down entirely; they serve more or less as a protection. It is well to treat them like the lily-stems, and not use the knife on many subjects, except gradually, as the stalks die down.

The coloring fruits, and the colored berries of many of the shrubs look so handsome at this season they might almost take the place of flowers. I question if the dahlia, in all its glory, can compare with many of the American crab-apples, or the double *Helianthus* hold more of yellow gold than the quince-trees are coining. The colored berries belong more truly to October than to September; they supply us largely with brilliant reds, a color the garden falls somewhat short of during the autumn, yellows being the dominant hues. The tall-growing *Helianthus orgyalis*, the fathom-high sunflower, is a late arrival—a dark-disked, golden-yellow flower, that looks down upon many of its tribe. One never knows at what elevation it will cease ascending until its sprays of blossoms unfold in late September. It is a lively, medium-sized flower, with delicate, long and very narrow lanceolate leaves, possessing what no others of the *Helianthea* with which I am familiar possess—a pleasant perfume. Its stalks, however, are rather feeble—they have so high to reach—and its effect is much enhanced by careful staking. It is best placed in the shrubbery, where it can receive partial support from other subjects.

Formerly the dahlia was much more frequently seen than now. Of late it is again becoming a favorite, many good singles and semi-doubles having been added to its numerous forms. The dahlia is a handsome old-fashioned flower, always effective in the shrubbery, and extremely desirable for cutting when arranged with its own foliage. Symbolically it stands for elegance and dignity. It might equally well be the type of steadfastness, the cut flowers being so lasting. None of its classes, including edged, tipped, laced, show, and fancy, are prettier than the tall, pure white, red, and yellow pompones. Many other forms are showier, with much larger flowers. The dahlia must be included among those flowers possessing a special scale of color, notably its dark reds, merging from vermilion into deep maroon, and folded petals almost like black velvet.

There are very many fine forms and varieties of the *Canna*, small and large, and with light- and dark-colored foliage. Principally planted for its grand foliage effect, its brilliant strelitzia-like flowers are highly ornamental as well. The *Canna* may almost supply the place of both *Tritoma* and gladiolus, combining, as it does, intense color of flower with great beauty of foliage and habit. So many species and varieties exist that a list of them would become tedious and confusing; the wild forms number nearly a hundred, with garden hybrids and seminal varieties innumerable.

The big phloxes continue to be the most magnificent flower of late September, the pure white " Vierge Marie" and the coppery-red " Oberon " showing superbly against a line of light-green and purple-leaved *Cannas*. Above the copse the sinking sun shines on the grand flower-heads of a long row, and earlier and earlier every day lights up the red trusses with intenser fire. The position is a partially shaded one, the soil light sandy loam well enriched, the plants five years old; in this position the grubs have scarcely troubled them. The phlox exhales a delicate yet pronounced odor, the sweetest-smelling flower of late autumn after the auratums have passed, except the fragrant double ten-weeks' stocks (*Matthiola annua*), the nutty odor of which it much resembles. The floral opposite of the crocus, the hardy colchicum or autumnal crocus, is now in bloom, its brilliant purple appearing after the leaves have died down, reversing the order of the spring flower it resembles. Or is the colchicum really the first spring flower, appearing months before the appointed time of its sisterhood?

Hydrangea paniculata grandiflora should be a conspicuous flower in all gardens during the autumn months. Unfortunately, it does not thrive in some soils, where it becomes a prey to the red spider. This species may sometimes be seen in perfection in one garden and worthless in an adjoining one. Apart from climatic influences the failure of certain plants is often puzzling. Much, I think, depends on vigorous subjects to start with. Many plants grown year after year by the nurseries in the same soil seem to become enfeebled, or at least to transmit a feeble habit to their offspring.

Situation likewise has much to do with the failure of a plant—too much sun, too much shade, or too much wind. Manure is frequently injurious to many plants, and grubs and insects are more numerous in some places than others. Some soils dry out quickly; others lack some essential element; still others become weak and deficient in vitality. Manuring in the latter case may assist but does not remedy the trouble. Working over the soil by deep trenching, and adding virgin soil and other elements that are wanting, is perhaps the most effective and troublesome way out of the difficulty. If one could only change old soil into new and transform an inland climate into a climate of the sea-coast at will, how much easier gardening would be! But even then there would be too much or too little lime, or something else would be wanting, I suppose—"man never is, but always to be blest."

Pope was a gardener, of course. That he was passionately fond of gardening can not be doubted in view of his statement, as given by Walpole, that of all his works he was most proud of his garden. He was a landscape-gardener rather than a floriculturist, however, painting with trees instead of flowers; and when we look over the great field of those artists whose canvas was Nature herself, where shall we find one who possessed the flowing, natural touch of Downing?

A wild garden, what Bacon termed his "heath or desert, framed as much as may be to a natural wildness," is a delightful feature of a place. Here among autumnal flowers belong many of the huge *Helianthea*, *Silphiums*, large starworts, and everlastings. There being no formality, it does not matter so much if the plants become overcrowded. The occasional presence of golden-rod will scarcely prove an intrusion, and the wild rose, bitter-sweet, fireweed, and clematis may be allowed to roam at will. Such a tangle should, of course, be placed in the distance, or spring upon one unawares. We should see more of this "natural wildness" in places whose extent and natural features are adapted to it, a source of far greater satisfaction than the flaring General Grant geranium-beds that often disturb the sense of repose. The prim modern garden, too, almost always lacks a pleasing feature of the ancient garden when rightly carried out; it has so few spots to lounge in. There is a dearth of garden-seats, niches, and benches, and vine-draped arbors and cloistered summer-houses. And where has the old sun-dial disappeared, that used to count the time so leisurely and shadow the passing hours?

The equinoctial has come and passed, shedding a mild persistent rain instead of the frigid down-pour it often brings. It cleared off with a blazing fire in a cool western sky and a mellow orange after-glow. A rarer, more exhilarating air has followed in its train, through which the first yellow streamers of the elms and birches gleam like molten gold. There is a richer color on the great hydrangea's plumes, a more satiny whiteness in the chaste blossoms of the larger anemone. A few days more, and the first white frost will settle upon the lowlands—a white mist rather than a white frost, that must soon set its blighting touch upon the flowers. If we might only store these golden autumn days to draw from during the tedious months of winter when the shadow is not on the dial! The next best thing is a gracious autumn lingering into late November, when the fire of the year goes out so slowly that it seems still to flicker amid the pattering rime.

Autumn is the harvest of those flowerless plants *par excellence*, the fungi, when old pastures and orchards and close-cropped sheep-walks yield up their treasures. There can be nothing pleasanter at this season than an expedition into the country in quest of the pink mushroom of the pastures. To inhale the air is in itself an inspiration, while road-side and lane are brilliant with the fall flora, and grasshoppers and crickets are chanting merrily in the fields. Then there is the excitement of pursuit and the triumph of capture. Mushrooms are like trout and game—they possess thrice the flavor where you earn them yourself or where they are sent by a friend. Neither should be purchased in the market; the bloom has been brushed off, the freshness has fled. The mushroom is more easily procured from the pastures than by artificial open-air culture. Like the poet, the out-of-door culturist is born, not made, and, I believe, must be born an Englishman. The leathery, insipid buttons the French hatch out and send us from their Cimmerian caves become worse and worse every year, like the cuts off the tough Montana rangers they are used to garnish. The large, high-flavored *Cep* of southern France, gathered like our *Agaricus campestris* from the open fields, is quite another thing, and is a prince of esculents when prepared *à la Bordelaise*. This agaric is very little known; it has even been overlooked by Gouffé and Francatelli. It may be had in good condition in the can of commerce, and, unlike the *champignon*, is always tender and digestible.

The common field mushroom itself is excellent *à la Bordelaise*, and, for those who do not know it, the recipe is worth quoting as a fragrant flower of the table when executed by a competent hand. There is no more expert guide than fat old Baron Brisse (*requiescat in pace*). I know of none so concise and explicit as the author of the " Petite Cuisine ":

"*Champignons à la Bordelaise.* — Choose large and freshly gathered mushrooms; wash, peel, and dry them; soak them an hour and a half in fine olive-oil with salt and pepper; then place them on the grill and turn them. After cooking, dress them on a platter and sauce them with hot oil, to which add finely chopped parsley and young onions and a modicum (*filet*) of vinegar." Here, as is observed in the recipe of Morilles *à l'Italienne*, "the trouble is trifling and the succulence is extreme."

The crisp air of October piques the appetite, and with the advent of the mushroom season one may be excused from turning for a moment from the flowers to the flesh-pots. There is a freshness about the "Petite Cuisine" that is truly delightful. Most of the very numerous books devoted to French cookery are so elaborate as to be practically useless. Pierre Blot's is an exception, and did much to simplify many dishes of merit. Baron Brisse has gone still further and contributed a gastronomic harmony that deserves to be translated into every language. His touch is so light; his faults are so few. Brillat Savarin was a cook of acknowledged ability. His "Physiology of Taste," however, is a monograph on the merits and etiquette of gastronomy rather than a practical guide to the preparation of the dishes themselves. Quaintness and simplicity are one of the charms of the "Petite Cuisine," wherein a *menu* is given for every day in the year. It is almost Lamb dressed in white cap and *marmiton* who presides at the range. Spring comes to Baron Brisse not with the first primroses, but with the first peas, and autumn possesses no tinge of sadness so long as it ushers in the hunting season and the spoils of the covers.

"Green peas! green peas!" he exclaims. "Of all street cries there is none that from basement to mansard so unanimously rejoices the hearts of all those that hear it. Green peas! green peas! This is the true spring; this is one of its most adorable gifts!"

He speaks of pigeons which join to a touching size an adorable savor. There are sixty-two ways of cooking them, he adds. "Some day I will give them all." The lark, always a favorite in France when done to a turn, he pronounces "detestable when not sufficiently cooked. If cooked too much it is still worse." A certain cream whipped with strawberries, of which he gives the recipe, he declares has left him many delightful souvenirs. Sorely distressed is he to trace the origin of a favorite *entrée* — sweetbreads *à la gendarme*. "This grasping (*empoignant*) *noix de veau à la gendarme*, is it the product of a man of arms, cook at his leisure, or of a master named Gendarme? Will it ever be known? My researches in this respect have been in vain."

But while the baron is happy in an *entrée*, he is pre-eminent in a *pièce de résistance*. The fertility of his resources is nowhere better illustrated than in his *resurgam* of a leg of mutton. "A roast leg of mutton," he truly observes, "when it is perfect as to quality, properly hung, and properly cooked, is a gift from heaven; but one finds it thus so rarely. A large gigot once cooked," he continues, "is supposed by housekeepers to be useless thereafter unless served cold or stewed. These ladies are mistaken; it is easy to present a leg of mutton on the table twice in the same conditions of excellence, and as intact in appearance the second time as the first.

"*Gigot de mouton rôti réchauffé.*—The gigot having been served once, and carved horizontally from one side only, wrap it in a piece of buttered paper and place on the spit. When well heated, lay it on a platter upon a generous *purée* of potatoes, the carved portion underneath; moisten gigot and *purée* with a portion of its juice which has been kept in reserve and heated without boiling, and serve." So much of our happiness here below depends upon the cook and the gravy! As in gardening, so in cooking—"*Ce n'est pas sans peine qu'on gagne le ciel!*"

To return to our mushrooms. Quantities of edible species exist in the fields and woods throughout the summer and fall, if we but knew them. They go to waste through our inability to distinguish the false from the true. The Gaul would have trained hogs and dogs to find them; the Italian would subsist on them in the dried state during winter. The silver spoon is a good though not always a safe test to distinguish them; better are the sweet odor and flavor, often resembling those of the chestnut, which characterize many of the edible species. Still, fungi are dangerous playthings for those not thoroughly experienced in gills and *pilei.* It is perhaps better that we are restricted to the field mushroom, than which no native species is more delicious, and in identifying which it is almost impossible to be mistaken unless one be color-blind, and can not distinguish pink from orange or saffron.

Brightest of the autumn flowers to enliven the lanes and road-sides are the purple asters, with the ever-surging sea of golden-rods, the rambling, canary-colored toad-flax (*Linaria vulgaris*), and an occasional pale-yellow evening primrose. Fields, meadows, and pastures are hoary with everlastings, and everywhere wave the white corymbs of the wild carrot. Here and there a stony field is sentineled with mulleins, on whose spires the goldfinches have congregated.

Along the road-side the elder-berry's cymes have been transformed to clusters of shining black berries, and ripe scarlet fruit shines through the tarnished foliage of the thorns.

The asters are swarming with bumble-bees and butterflies—the small white and yellow butterflies and the larger orange-blacks, all busily extracting a "last taste of sweets." I did not know the latter was ever so late an arrival, or that his chrysalis so resembles a Japanese watch-charm. I found two belated chrysalides on the raspberry-vines. The color of the envelope was dark bronze. Near one extremity were two burnished silver knobs; near the other, a necklace of raised, shining gold and enameled beads. Underneath the semi-transparent envelope the folds of the orange wings showed. I found the thin husks rent in twain the following morning in the glass in which they were placed, the two perfected insects struggling to escape from their narrow confinement. Brief will be their holiday in the slant autumnal sunshine, and "too late" the burden borne to them by the rustling breeze. Last year there was a storm of these brilliant insects in a neighboring grove, where they settled so numerously as to weigh down the lesser limbs. The year previous, a similar occurrence was noticed along the lake-shore.

The pretty yellow flower you noticed a month since along the ditches and low places is scarcely recognizable now. A friend then, to all outward appearances, it has turned to a foe, thrusting its javelins at whatsoever crosses its path. The bright-yellow petals have disappeared, the green disk has changed to rusty brown; and the larger burr-marigold (*Bidens chrysanthemoides*) stands revealed in its true hideousness—an ugly, swarthy ruffian, at once an armory of halberds, arquebuses, arrows, and poniards.

Finest of the extremely numerous asters that follow one during an autumnal ramble are the several forms of *A. Novæ Angliæ*, the large purple starwort of the road-sides, varying from lavender and rosy purple to deep purple. It is the richest and one of the gayest of the common late wild flowers, and, common as it is, is well worth a suitable place in the garden. It seldom looks so well under cultivation, for the reason that it is seldom planted in sufficient quantity. This is equally the case with most wildlings; they should be seen in masses, as they occur naturally, to disclose their true worth. "Enough" is not "as good as a feast" when it comes to flowers.

The late flowering characteristic of the perennial aster is one of its many charms. Gérarde, in his "Great Herball," as long ago as 1633, refers at length to the North American aster, two species being described by him (p. 489), one of which, not previously mentioned, he states "is to be esteemed for that it floures in October and November, whenas few other floures are to be found." Certainly, many of our native species exceed in beauty the species mentioned in the fourth chapter of the " Georgics " (*Aster amellus*, the Italian starwort). The classic reference is altogether too pretty not to transscribe :

>Est etiam flos in pratis, cui nomen Amello
>Fecere agricolæ, facilis quærentibus herba.
>Namque uno ingentem tollit de cespite silvam,
>Aureus ipse ; sed in foliis, quæ plurima circum
>Fundunter, violæ sublucet purpura nigræ.
>
>In Meades there is a flower Amello named,
> By him that seeks it easy to be found,
>For that it seems by many branches fram'd
> Into a little Wood ; like gold the ground
>Thereof appears, but leaves that it beset
>Shine in the colours of the Violet.

So highly was the aster's purple and gold esteemed that the flowers were used as offerings in religious rites, as Virgil specifies in his further reference to the Amello.

Cheerful and colorful are the annual asters if

the strain be choice and the soil an unctuous loam. Sown late in May, so the plants may come into blossom during latter September and early October, they are seen at their best. It needs cool weather with just a suspicion of white frost to bring out their colors. Then, when many of the perennials are in the sere and yellow leaf, they lend an almost spring-time gayety to the garden. But here, as with the *Pyrethrums*, objectionable shades must be banished, and the whites, maroons, soft roses, lilacs, lavenders, and purples, placed so they may harmonize and their various lights may shine. They are the roses of autumn, the more beautiful because their reign is so fleeting. But the annual aster is invariably sown too soon.

Strikingly beautiful are the calendulas during October. Daring in their hues as the zinnea, they never overstep the limits, and do not attempt to mix up crimsons with yellows. The orange verging to red, and the gradual shadings from buff to yellow and salmon of the rays, are a study and a joy in color. They last so long, and withstand the frost so bravely, that the rear garden and the center-table would seem lonely without them, and we may freely forgive their somewhat acrid odor. I found a large bunch of them upon the table to-day, in a low, blue *cloisonné* vase, the slanting afternoon sun streaming full upon them—an October sunset in the room. There should be a shelf of vases to choose from for arranging flowers—tall, flat, large, and small; the floral picture, too, calls for its appropriate frame.

Helianthus tuberosus, the Jerusalem artichoke, shows a fine mass of yellow far above one's head, an erect, vigorous grower, with large, dark-green leaves and lively flowers. In its habit, and the size and brilliancy of its blossoms, its surpasses *H. giganteus*. It comes late into blossom, and defies the frost. This, with many of the taller species, as has been stated before, looks best in the distance naturalized in large masses. They are admirably suited to low situations, where they can be looked down upon from an elevation. The tubers of *H. tuberosus* would be largely used as an esculent, if we had not the potato. They have a flavor somewhat like salsify or celery-turnip, and it seems highly probable that they were extensively employed by the aborigines as an article of food.

Very brilliant are the shining berries of many of the ornamental shrubs at this season. The large fruit of the Cornelian cherry (*Cornus mascula*), and the corals of the bush cranberry

(*Viburnum opulus*), still hold their color. Gay scarlet streamers wave from the common barberry bushes. The large flat cymes of the wayfaring-tree (*V. lantanoides*) are covered with red drupes, changing to dark-purple, while the foamy blossoms of the rough-leaved *Viburnum* (*V. rugosum*) have been succeeded by the showy berries of fall. Indeed, these shrubs have been brilliant with fruit ever since latter August.

I had not meant to overlook the baneberry, attractive in May with its spiræa-like flowers, and bright throughout August and September through its ovoid, oblong, red berries. Two varieties are worth cultivating — *Actæa alba*, having white berries with red stalks, and *A. spicata rubra*, bearing glossy vermilion fruit. But the baneberry requires partial or nearly entire shade, and plenty of moisture, or the leaves soon tarnish, and give the plant a withered appearance. It looks well suitably placed in the rock-garden, or rising from the shrubbery border, or the wild garden. Nearly allied to the baneberry is the black snake-root (*Cimicifuga racemosa*). You have noticed it when passing along woods on the railway, lighting their green skirts with its tall white rockets of bloom. An odd, bold, and distinct flower, it is not unworthy of culture despite the unpleasant odor, whence the species derives its name, from *cimex*, a bug, and *fugo*, to drive away.

The vermilion and light-red berries of the European and the American mountain-ash (*Pyrus aucuparia* and *P. Americana*) are conspicuous at present. One of the finest ornamental trees, the mountain-ash, like the linden, is unfortunately subject to attacks from borers and ants, which eventually split the rind and destroy the vitality of the tree. The robins are numerous among its branches, feasting upon the berries. Equally busy are they among the pears, apples, and grapes—meat-eaters in the summer and vegetarians in the fall. The robin has a distinct autumnal note, which I like to hear—a noisy call he utters when about to change his perch, or preliminary challenge to a raid upon the orchards, as if he knew he had a right to the spoils, and wanted his companions to share the feast. From the thickly foliaged thorn, hung with its scarlet fruit, comes a soft, tender, caressing song, one of the sweetest of the year—a warble so low, so sweet, so plaintive, I tiptoe closely to the songster to hear it. How charming the cat-bird can be when he tries, and how different his dulcet autumnal vespers from the frenzied "Czardas" he is so fond of playing in the morning of the year!

I know many a man like him—grouty, fault-finding, storming in the morning; mellow, expansive, delightful in the evening.

Now crickets crink by day, and the harping of grasshoppers ascends from the fields. Countless unseen choristers are chanting an ode to fall—the air quivers with pulsating sound. Even throughout the October night the viol of the green leaf-cricket is never stilled. Musically the squirrel's bark rings out from the covert, and intermittingly rises the warbling of assembling blackbirds. Over the flowers swarm crowds of sulphur butterflies, and bee and wasp are banqueting upon the fallen fruit. Perceptibly the shadows lengthen, as the haze of autumn draws its veil over the latter year. Soon, ah! why alway so soon? the patter of dropping nuts and the rustle of falling leaves.

Every little while I catch a fragment of a familiar strain voiced by the song-birds on their southward flight as they pause for a day on their migration. From what distant coverts and unexplored forests has not that white-throated sparrow returned, whose silvery tinkle floats from the copse so musically, yet so plaintively, seeming like an echo of departed spring!

The yellow-birds, who are busy scattering the milkweed's floss, have a little lisping cry that always seems tinged with sadness at this season. Perhaps the season has more to do with the apparent sadness than the voice of the bird itself. If the frogs were vocal in October, no doubt the trombone of the great green batrachian would seem a *Miserere*. Were the green leaf-cricket a spring chorister, his measured "Treat-treat-treat" would doubtless appear a buoyant "Frühlingslied." So much depends on association of familiar sounds with the season, or the circumstances under which they are heard. I can scarcely imagine how the call of the meadow-lark would sound from the depths of a thicket, or how much of its metallic quality the veery's song would lose if uttered in the open field.

But the blackbird's notes during autumn are assuredly sad, as they linger over the withering stubbles, or drop down from the home-bound flocks at evening. Every morning, now, they pass overhead in large bands from the marshes, on the way to their daily forage-grounds; and every evening, now flying low and now flying high, they return over the self-same route to the haven of the reeds. The majority are blackbirds, though the starling and crow-blackbird feed with them, and form part of the morning and evening flights. The flocks grow larger as the season advances, and, when flying low in the calm of evening, cause a sishing sound, like the ebb of the surf upon the shingle. What a clamor there arises from the ebon flocks in the corn fields for weeks before their departure; what garrulous sessions are held by the disputing crowds ere the date is fixed upon for their southward flight! We may well wonder how the young birds are made to understand the signal of departure, and marvel

> Who calls the council, states the certain day?
> Who forms the phalanx, and who points the way?

The Last Monk's-hood Spire.

For never-resting Time leads Summer on
  To hideous Winter, and confounds him there;
Sap check'd with frost, and lusty leaves quite gone.
  Beauty o'ersnow'd, and bareness everywhere.
                                    SONNET V.

XV.

THE LAST MONK'S-HOOD SPIRE.

HERE is little left to tell of the flower-garden after mid-October; its brightness fades rapidly with the shortening days. Glorious have been the great Japanese anemones; they are the life of the borders in October, being to autumn what the daffodils are to spring. Dahlias, salvias, and ageratums have been struck by frost; the anemones still linger, white as the snow-flakes they herald. The chrysanthemums are just appearing, among the latest of autumn flowers, and we once more touch our hat to China and Japan. For the fading flowers we have the brilliant fruits and berries, and the changing hues of the foliage. On yonder upland grove I see nearly every shade of red and yellow which the entire summer has contributed to the flower-borders. The maple would be held a sacred tree by the Orientals for its skill as a fall landscape-painter. Almost equally beautiful is the dogwood, a tree that should be in every garden, no less for its October splendor than for its magnificent June inflorescence.

It is an opportune moment to consider trees and shrubs with regard to their autumnal hues; later they may be studied with reference to the beauty of their spray and leafless lines. The feature of autumnal coloring should receive attention when planting, just as much as the flowering habit of ornamental trees and shrubs.

The very high coloring of foliage in nature we may not always hope to equal, for the reason that intensity of hue is frequently caused by overcrowding, poor soil, or special exposure, the latter being undoubtedly the most important factor. Trees growing on arid and stony ground are usually the most highly colored, though frequently a swamp, where the scarlet maple and sweet-gum flourish, glows like a lambent flame. Full maturity of foliage before it is touched by black frosts, and position with reference to the sun, also count for much in the bursts of color that hang upon the autumnal upland.

Many kinds of trees and shrubs assume the same hue or hues every autumn, individuals offering no deviation. Other kinds, like the scarlet and sugar maples, are widely different in the colors individual trees assume. It may be noticed that a tree with individual markings always repeats these markings—the same red branch or branches, or the same scarlet leaves tipped with green, duplicating themselves on the same tree year after year. It would be well if nurserymen would propagate, through grafting, striking individual trees for their autumnal coloring, notably the scarlet and sugar maple.

Without doubt the maple is king of arboreal colorists, no other tree presenting so great a variety of glowing hues. The scarlet maple contributes more self-colors than the sugar. Its leaf, however, falls quicker, and does not possess the delicate shadings from green to reds and yellows that many of the sugar-maples do. But in its first flush of scarlet, orange, or cardinal it has no rival for distant effects. The larger Japanese maple (*Acer polymorphum*) should always have a place, for the beauty of its autumnal foliage. Indeed, the smaller Japanese maples, as well, are of marked beauty during autumn. Among trees that assume a bright yellow, the Norway maple, elm, birch, hickory, maiden-hair, ash, yellow-wood, and larch are conspicuous. The sassafras has its individual hues—ochres, passing from yellow to deep orange and umber. The shad-blow colors a rich garnet, not unlike some of the tones of the pepperidge-tree; while the dogwood's is unquestionably the most vivid, deep lake-red of all trees. The American mountain-ash passes from yellow to rich clarets and purples; the European mountain-ash seldom develops much autumnal coloring, confining its display to its brilliant fruit. The sweet-gum and sour-gum, fine trees at all seasons, are exceptionally attractive during autumn in the deep purplish-red and orange shades of the leaves.

For simple variety of colors, the various oaks are almost equal to the maples. The oak has its own scale of russets and maroons; and no one can pass it without admiration, when the November sunlight strikes through the glistening foliage of the native scarlet oak, the last bright-red of fall.

If we take yellow alone for the color-standard, the beech is without an equal. A beech, indeed, is always beautiful. In late November its colors still remain attractive, varying from rich Roman ochre to deep-brown bronze, and from pale rose-buff to lustrous, satiny gray. Assuredly Downing is mistaken in considering its beauty diminished during winter, owing to the retention of much of its foliage. Its harmony is of marked loveliness in winter, a faded elegance clinging to it like a chastened autumnal memory. I can not understand how Wilson Flagg should refer to it as remarkably dull in its autumnal tints. To the Selborne rector the beech was "the most beautiful of all trees," and Jesse rightly "loved it at all seasons of the year."

Among smaller trees, the aspen is prominent for its golden-yellow hue, its effect being heightened by the play of the sunlight upon its quivering leaves. The common sumac is invariably one of the most brilliant colorists, especially when growing on stony places. The cut-leaved variety (*Rhus glabra laciniata*), a striking shrub, with deeply-cut, fern-like foliage, is equally beautiful in its October dress. Several of the shrubby spiræas are worth planting solely for their autumnal foliage, particularly the plum-leaved variety (*S. prunifolia*). But, of all small ornamental shrubs, the finest is *Berberis Fortuneii*, the small leaves of which vary through different shades of green, yellow, and salmon to vivid Venetian red. Of fruit-trees that contribute to the autumnal pageant, the most striking are the peach, pear, apple, and cherry. In the two former, greens are often most exquisitely graduated, passing into yellow, orange, and red; the apple preserves its green for a long period, and then, in numerous varieties, shades it with yellow before the leaves become seared by hard frosts.

It is self-evident that there can be no satisfactory garden without a sufficiency of trees and shrubs. The former are necessary, if only for shade. But trees and shrubs with colored and variegated foliage, and those which assume vivid autumnal tints, are rarely seen as frequently under cultivation as they should be; and many a garden, for this reason, lacks a great charm of outward nature.

For several reasons, fall is the best season for transplanting. One can judge better, at least so far as the shrubberies and flower-borders are concerned, *where* to plant, than when the plants have died down, or are denuded of foliage. Moreover, when planting is deferred until spring, many things are apt to be forgotten in the rush of garden-work. The sooner you plant a desirable tree, shrub, or flower, the sooner you will derive the benefit. Even a fine specimen perennial often requires years to attain its development. The proper way, it may be reiterated, is to plant something every year; and it is better to plant excessively, thinning out as becomes necessary, than to plant sparingly.

The older you grow, if you love your garden, the more your taste will develop, and the more you will regret not having set out a tree, shrub, or perennial in the place it might occupy and adorn.

Autumn is variously voiced by the poets, more often in a minor than a major key. Despite the pomp with which she appears, her crimsoning woods are but the presage of approaching death, when the snow shall be her burial shroud and winter's winds shall chant her funeral dirge. Charming she is in her mingling of October sunshine and shadow; pitiful in her mournful November garb. Yet let but a burst of sunlight touch the leafless trees, and she is instantly transformed.

In British verse autumn is usually dank and sodden, bleak or shivering. The yew and the holly seem to absorb the light and cast a pall upon the landscape. The sugar and scarlet maple, the dogwood and sumac, are wanting to impart their warmth of color; and St. Martin's summer somehow fails to shed a cheerful influence as does our Indian summer. Thus David Gray:

> October's gold is dim—the forests rot,
> The weary rain falls ceaseless—while the day
> Is wrapped in damp. In mire of village way
> The hedgerow leaves are stamped; and, all forgot,

The broodless nest sits visible in the thorn.
Autumn, among her drooping marigolds
Keeps all her garnered sheaves, and empty folds,
And dripping orchards—plundered and forlorn.

Even Shakespeare shivers:

That time of year . . .
  When yellow leaves, or none, or few, do hang
Upon those boughs which shake against the cold,
  Bare ruin'd choirs, where late the sweet birds sang.

Tennyson is pathetic, but neither somber nor gelid:

Calm is the morn without a sound,
  Calm as to suit a calmer grief,
  And only through the faded leaf
The chestnut pattering to the ground:

Calm and deep peace on this high wold,
  And on these dews that drench the furze,
  And all the silvery gossamers
That twinkle into green and gold.

Of all odes to autumn, Keats's, I believe, is most universally admired. This might almost answer to our own fall of the leaf, and is far less somber than many apostrophes to the season that occur throughout English verse. Another contemporaneous ode, though less generally admired, is, I think, equally fine and certainly stronger. Hood's is emphatically an ode to late

November; Keats's applies more strictly to late October. Each is perfect in its way. Between them exists the same difference as there exists between Keats's and Leigh Hunt's rival sonnets to the grasshopper and cricket. Keats's is less forceful. Could there be anything stronger than Hood's grand opening lines?

> I saw old Autumn in the misty morn
>   Stand shadowless like Silence listening
>   To silence, for no lonely bird would sing
> Into his hollow ear from woods forlorn,
> Nor lowly hedge, nor solitary thorn.

It is the very shadow of November, when the fire of autumn is burned out, and shivering Nature silently awaits the shroud which is to cover her. These four lines have rarely been equaled in the picture they convey of autumn desolation:

> Where is the pride of summer—the green prime—
>   The many, many leaves all twinkling? Three
> On the moss'd elm, three on the naked lime
>   Trembling, and one upon the old oak-tree!

And again:

> The squirrel gloats o'er his accomplish'd hoard,
> The ants have brimm'd their garners with ripe grain,
>   And honey-bees have stored
> The sweets of summer in their luscious cells;
> The swallows all have wing'd across the main;

> But here the Autumn melancholy dwells
>   And sighs her tearful spells
> Among the sunless shadows of the plain.
>   Alone, alone,
>   Upon a mossy stone,
> She sits and reckons up the dead and gone,
> With the last leaves for a love rosary. . . .

Keats's ode is less austere. It has more of autumn gold than maroon; more of purple haze than leaden skies. Thus, the second stanza:

> Who hath not seen thee oft amid thy store?
>   Sometimes whoever seeks abroad may find
> Thee sitting careless on a granary floor,
>   Thy hair soft-lifted by the winnowing wind;
> Or on a half-reaped furrow sound asleep,
>   Drowsed with the fume of poppies, while thy hook
> Spares the next swath and all its twinèd flowers:
>   And sometimes like a gleaner thou dost keep
> Steady thy laden head across a brook;
> Or by a cider-press, with patient look,
>   Thou watchest the last oozings hours by hours.

The alliteration in *s* is noticeable in each of the two preceding stanzas; but Hood's felicitous use of the vowel *o* throughout the ode imparts to it a solemnity and gloom that express the mournful spirit of November such as has no counterpart in poetry inspired by the latter season.

Thomson's old etching of Autumn still stands out as sharply as when first defined:

> Crown'd with the sickle and the wheaten sheaf,
> While Autumn, nodding o'er the yellow plain,
> Comes jovial on.

Its expressiveness must have caught the fancy of the French, for a Gallic couplet reads:

> Couronnée d'épis, tenant en main la faucille,
> L'Automne joyeuse descend sur nos campagnes jaunissantes—

which, if not a literal transcription, bears its coloring in a marked degree.

Herrick paints Autumn as

> The Northern Plunderer
> To strip the Trees and Fields to their distresse,
> Leaving them to a pittied nakednesse.

I have always admired a version of Autumn by an old master who painted in prose:

> Autumn is the barber of the year who shears the bushes, hedges, and trees—the ragged prodigal who consumes all and leaves himself nothing; and this bald-pated Autumn is seen going up and down orchards and groves, fields, parks, and pastures, shaking off fruit and beating leaves from the trees.

Charles Tennyson Turner's "October," like all his sonnets, is stamped with a delicate and graceful fancy:

> 'Twas the last week the swallow would remain.
> How jealously I watched his circling play!
> A few brief hours and he would dart away,
> No more to turn upon himself again.

A more tender melancholy pervades the companion sonnet to "Autumn":

> The crush of leaves is heard beneath his feet,
> Mixt, as he onward goes, with softer sound,
> As though his heel were sinking into snows.
> Full soon a sadder landscape opens round,
> With, here and there, a latter-flowering rose,
> Child of the summer hours, though blooming here
> Far down the vista of the fading year.

The sounds of latter autumn, which we have all listened to from some still upland, are articulate in "An Autumn Landscape," by Alfred Billings Street:

> Far sounds melt mellow on the ear; the bark—
> The bleat—the tinkle—whistle—blast of horn—
> The rattle of the wagon-wheel—the low—
> The fowler's shot—the twitter of the bird.

Nowhere in American poetry, however, are the lights and shadows of Indian summer drawn with a truer touch than by Lowell and Read. Thus the former's "An Indian Summer Reverie":

> Far distant sounds the hidden chickadee
>   Close at my side; far distant sound the leaves;
> The fields seem fields of dream. . . .

The cock's shrill trump that tells of scattered corn,
  Passed breezily on by all his flapping mates,
Faint and more faint, from barn to barn is borne,
  Southward, perhaps to far Magellan's straits. . . .

The single crow a single caw lets fall;
  And all around me every bush and tree
Says Autumn's here and Winter soon will be,
  Who snows his soft white sleep and silence over all.

And Read, in "The Closing Scene":

All sights were mellowed and all sounds subdued,
  The hills seemed farther and the streams sang low;
As in a dream the distant woodman hewed
  His winter log with many a muffled blow. . . .

The sentinel cock upon the hill-side crew,
  Crew thrice, and all was stiller than before—
Silent till some replying warder blew
  His alien horn, and then was heard no more. . . .

But, amid the melancholy of the autumn Muse and the gloom of autumnal skies, I catch a pleasing fancy to nurse through the tedious winter hours. I thought the crocus the herald of spring; but in the copse I already catch a gleam of vernal gold. The witch-hazel (*Hamamelis Virginiana*) is first to put forth its sturdy blossoms, pure and fresh at this season as was the gilded urn of March. Often.I meet its flower-clusters in the wintry woods when all its companions save the oak, beech, and hornbeam are denuded of foliage, smiling at the cold and snow. Does it not convey a meaning? Its pale-yellow petals speak to me of immortality, and its fragrant breath exhales a promise of coming flowers.

What more remains to say of the garden, now shorn of its beauty, except that each year one learns to love it more? Alone, defying frost and sleet, the tall blue monk's-hood spires remain, to be stricken down in turn, and patiently await the dawn of spring.

Index

Acer polymorphum, 327.
  rubrum, 27, 28, 326, 327.
Achillea rosea, 51.
Actæa, 318.
Actinidia polygamia, 241.
Adder's tongue, 71.
Addison on the fancy, 7.
Adiantum pedatum, 268.
Adonis vernalis, 131.
Agaricus campestris, 308.
Ageratums, 325.
Akebia quinata, 242.
Allen, Grant, 206, 210.
Alpine catchfly. See SILENE ALPESTRIS.
Althæas, 35, 36, 234, 236, 276.
Alyssum, 131.
Amelanchier Canadensis, 73.
Androsace, 131.
Anemone, 110, 123, 124, 127.
  Alpina, 123.
  blanda, 123.
  fulgens, 123.
  Japanese, 37, 52, 280, 306, 325.
  nemorosa, 123.
  palmata, 123.

Anemone pulsatilla, 123.
  sylvestris, 123.
Angling masters, ancient, 201.
Anthemis tinctoria, 160, 247.
Ants, 45, 199, 319.
Aphides, 204.
Apios tuberosa, 242.
Apple-blossoms, 86.
Aquilegia, 141.
  Canadensis, 72, 142.
  chrysantha, 141.
  cœrulea, 141.
  glandulosa, 142.
  longissima, 141.
  Stuarti, 142.
  vulgaris, 142.
  Witmanni, 142.
Arbutus, 5, 61, 63, 110.
Arethusa bulbosa, 129.
Arisæma triphyllum, 72.
Aristotle, 214, 250.
Arum arisarum, 73.
Ascham, Roger, 30, 31.
Asclepias tuberosa, 251.
Ash, 327.
Aspen, 198, 329.
Aspidium achrosticoides, 267.

Aspidium aculeatum, 267.
  cristatum, 267.
  felix-mas, 267.
  Goldianum, 267.
  marginale, 267.
  spinulosum, 267.
Asplenium ebeneum, 270.
  nigrum, 270.
  trichomanes, 270.
Association in odors, 7.
Aster amellus, 315.
Aster, annual, 315.
Aster Novæ Angliæ, 314.
Asters, 5, 256, 312–314.
Autumnal hues, 306, 326.
Autumn and the poets, 331.
Autumn, the ode to, 320.
Avens, scarlet, 248.
Azalea, 110, 139, 140.
  calendulacea, 140.
  nudiflora, 140, 255.
  viscosa, 140, 255.

Bachelor-buttons, 146, 250.
Bacon, Francis, 33, 34, 305.
Baneberry, 318.
Barberry, 35, 236, 318.
  purple-leaved, 235.
Bats, 149, 221, 222.
Baudelaire, 7.
Beeches, 255, 328, 337.
Beech-fern, 78, 269.
Bee-larkspur, 210–212.
Bee-masters, ancient, 199.
Bees, 106, 159, 188, 193, 204, 209, 212, 216, 218, 219, 230, 241, 291.

Bees, a singular preference shown by, 210.
  colors preferred by, 209.
  perforating the corolla, 212.
  and bears, 200.
Beetles, destructive to flowers, 217.
Bell, Peter, 54.
Belleau, Remy, 59.
Bell-flower, 146, 156, 157.
Bellworts, 73.
Berberis Fortuneii, 329.
Bermuda lily, 176.
Bidens chrysanthemoides, 314.
Bills, the abomination of, 197.
Birds, migrating, 320.
Bitter-sweet, 305.
Bitter-vetch, 160.
Blackbird, 18, 81, 320, 321.
Black-cap chickadee, 255.
Bladder-fern, 77.
Bladder-senna, 235.
Blood-root, 59, 63, 110.
Blot, Pierre, 309.
Bluebird, 18, 27, 59.
Blue jay, 20, 253.
  flag, 72.
Bluets, 78.
Bobolink, 82.
Bocconia cordata, 284.
Bog-garden, the, 33, 76.
Bokhara bell-flower, 157.
Bombus terrestris, 212, 219.
Books I must read, 25.
  for summer reading, 193, 202.

Books, pocket editions of, 87.
Borders, flower, 46, 51, 52.
  size of, 38.
Borers, 199, 319.
Bosoms of the beautiful, 155.
Botanists, the German, 213.
Botrychium Virginicum, 267.
Bowne, Borden, 24.
Bracken, 267.
Brantôme on feminine beauty, 67.
Brier rose, 146, 189.
Brisse, Baron, 10, 308, 309.
Brown-creeper, 255.
Bruchus pisi, 21.
Bryant, 17, 65.
Bull-frogs, 105, 321.
Bulwer, 8, 194, 228.
Bumble-bees, 159, 202, 204, 209–213, 219, 313.
Buphthalmum cordifolium, 247.
Burbidge, F. W., 79, 91.
Burns, Robert, 120.
Burr-marigold, 314.
Burroughs, John, 16, 19, 87, 277.
Bush cranberry, 317.
Butcher-bird, 19.
Butler, 200.
Butterflies, 193, 219, 223, 313.
  a storm of, 313.
Butterfly-weed, 251.
Button-bush, 233.

Calendula, 156, 276, 316.
Californian lilies, 175.

Calla-lily, 5.
Caltha palustris, 75, 76.
Calycanthus, 35, 235.
Camellia, 5.
Camomile, 160.
Campanula barbata, 157.
  Carpatica, 300.
  macrantha, 157.
  medium, 156.
  persiscifolia, 157.
  pulla, 157.
Camptosorus rhizophyllus, 270.
Canna, 300, 302.
Canterbury bells, 156.
Cardinal-flower, 110, 136, 183, 252.
Carnation, 155, 250.
Castle of Indolence, the, 193.
Catalogues, the florist's, 14, 88, 295.
Cat-bird, 82, 159, 319.
Celastrus scandens, 242.
Centaurea dealbata, 248.
  glastifolia, 247.
  macrocephala, 247.
  montana, 247.
  moschata, 248.
  Ruthentica, 247.
Cephalanthus occidentalis, 233.
Ceterach officinarum, 270.
Cheilanthes vestita, 270.
Chelone glabra, 252.
Cherry, double-flowering, 86.
Chestnut-tree, 250.
Chickadee, 255, 336.

Chionanthus Virginica, 230.
Chionodoxa Lucilæ, 107.
Chrysanthemum, 325.
  maximum, 246.
Cicada, 106, 250, 278, 280.
Cimicifuga racemosa, 318.
Cineraria, 51.
Cinquefoils, 131.
Cladrastis tinctoria, 230.
Clare, John, 289.
Claytonia parviflora, 63.
  Virginica, 62.
Clematis, 237-239, 242, 305.
  Jackmani, 34, 237.
Clethra alnifolia, 74, 232, 233.
Climbing shrubs, 236.
  staff-tree, 242.
Club-moss, 255.
Colchicum autumnale, 303.
  lily, 170.
Collins, William, 70.
Colors, 5, 8, 15, 38, 48, 55, 61, 76, 121, 138, 139, 150, 152, 153, 174, 198, 209, 224, 235, 245, 246, 252, 257, 276, 281, 301, 302, 306, 312, 316, 317, 318, 325-330.
  autumnal, 300.
  unhappy use of, 153.
  when they harmonize, 153.
Columbines, 52, 72, 141.
Colutea, 235.
Compass-plant, 293.
Coptis trifolia, 127.
Corchorus, 236.
Coreopsis, 156, 295.
  lanceolata, 161, 247, 295.

Cornelian cherry, 236, 317.
Cornus Canadensis, 77, 78.
  floridus, 74, 326, 328, 331.
  mascula, 317.
Corolla, perforated by bees, 212.
Cotoneaster, 236.
Country gardens, 53, 143, 145, 156, 181.
Cowslips, 4, 106.
Crab, garland-flowering, 86.
Cranesbill, 72.
Crickets, 7, 106, 234, 251, 278, 300, 307, 320.
  climbing, 279.
  green leaf, 241, 251, 278-280, 320, 321.
Crocus, 28, 110, 303, 337.
  autumnal, 303.
Cross-fertilization, 215.
Crow-blackbird, 16, 20-22, 82, 321.
Crown-imperial, 61, 146.
Crows, 15, 206, 337.
Cryptogramme acrostichoides, 270.
Culpepper, 290.
Currant, yellow flowering, 36.
Cypripedium acaule, 128.
  arietinum, 129.
  parviflorum, 128.
  pubescens, 128.
  spectabile, 128, 183.
Cystopteris bulbifera, 268.
  fragilis, 77, 78, 268.
Czardas, a Hungarian, 82, 319.

Daffodils (see also NARCISSUS), 4, 5, 49, 52. Chapter IV, 109, 110, 146, 325.
catalogues, 88.
causes of color in, 95.
culture of, in England and Holland, 98, 99.
dance of the, 96, 187.
hoop petticoat, 90, 98.
Horsfieldi, 4, 87, 92.
hybridizing, 91, 92.
innumerable forms of, 89.
Dahlia, 146, 301, 302, 325.
Daphne blagyana, 118.
cneorum, 3, 117.
mezereum, 3, 36.
rupestris, 118.
Darwin, Charles, 209, 212, 215, 224.
Dawn, a summer, 159.
Day-lilies, 161.
white, 146, 161, 285.
yellow. See HEMEROCALLIS FLAVA.
Dead March in Saul, the, 197.
Délille, Jacques, 101, 104.
Delphinium Wheelerii, 210–212.
Deutzia, 35, 36, 229.
crenata fl. pl., 229.
gracilis, 229.
pride of Rochester, 229.
scabra, 229.
Dianthus, 131, 155.
plumarius, 156.
Dicentra cucullaria, 72.
Dictamnus fraxinella, 145.

Dictionary, the, illustrations in, 25.
Dielytra, 146.
Diervilla. See WEIGELA.
Dioscorides, 214.
Dobson, Austin, 194.
Doctor, my allopathic, 24.
Dod, Rev. C. W., 66, 115, 285.
Dodecatheon, 131.
Dog-tooth violet, 59, 71.
Dogwood, 74, 316, 318, 321.
shrubby, 235, 236.
variegated, 36.
Double-flowering rocket, 146, 161.
Downing, A. J., 101, 305, 328.
Downy woodpecker, 255.
Dryden, John, 104.
Dumas, père, 9, 87, 197.
Dusk, a summer, 161.
Dutchman's breeches, 72.
Dwarf cornel, 77, 78.
spleenwort, 78.

East wind, the, 23, 203.
Eau de Cologne, 8.
Echinacea, 295.
Elæagnus edulis, 236.
Elder, 231, 313.
black-fruited, 236.
cut-leaved, 36, 232.
fern-leaved, 36, 232.
golden-leaved, 36, 231.
variegated-leaved, 231, 236.
Emerson, R. W., 110.
Epilobium, 256.

Equinox, 14, 23, 24, 306.
Erianthus ravennæ, 37.
Erlking, the, 197.
Erythronium Americanum, 71.
 giganteum, 71.
 grandiflorum, 71, 117.
 Hendersonii, 71.
Eschscholtzia, 156, 276.
Ettrick Shepherd, the, 16.
Eulalias, 286, 300.
 gracillina univitata, 300.
 Japonica, 37, 300.
Euonymus, red-fruited, 236.
 white-fruited, 236.
Eupatorium, 256.
Evening primrose, 98, 256, 286, 287, 312.
Ever-blooming pea, 161.
Evergreens, pruning, 37.
Everlasting, the field, 7, 136, 257, 305, 312.
Exochorda, 36, 235.

False miter-wort, 126.
 Solomon's-seal, 78, 125, 127.
 violet, 127.
Fathom-high sunflower, 301.
Ferns, 33, 198, 255, 261.
 beech, 78, 269.
 Christmas, 267.
 cinnamon, 266.
 evergreen wood, 267.
 hart's-tongue, 270.
 interrupted, 265.
 maiden-hair, 261.
 oak, 77, 78, 127, 269.
 ostrich, 185, 261, 264.

Ferns, polypody, 77, 78, 126, 127, 269.
 royal, 137, 241, 265.
 sensitive, 266.
 shield, 267.
 spleenwort, 270.
 walking, 270.
Fertilization of flowers, the, 214.
Fire-flies, 161.
 weed, 305.
Flagg, Wilson, 329.
Fleur-de-lis, 139.
Flora, spring and summer, compared, 135.
Floral millennium, advent of, 15.
Flower catalogues, 14, 34, 88.
 customs, Oriental, 151.
Flowers at the grave, 6.
 for cutting, 138, 153, 156, 302, 316.
 indoor, 6.
 nocturnal, 223, 224, 286, 287.
 old-fashioned, 145, 148, 155, 249, 302.
 prevailing colors of, 224.
Foliage, autumnal, 325-331.
 colors of spring, 61.
Forsythia, 36, 83, 84.
 Fortuneii, 83.
 suspensa, 83.
 viridissima, 83.
Fraxinella. See DICTAMNUS FRAXINELLA.
Fringed polygala, 77, 78.

Frogs, 27, 201, 321.
Frosts, the first, 306, 316, 325.
Funkia grandiflora, 285.
Fungi, 307, 312.

Garden, a cool place in the, 198.
 a neglected, 148.
 a reserve, 49.
 a wild, 305.
 a wild woods, 78.
 the bog, 33.
 the country, 53, 143, 145, 156, 181.
 the formal, 32.
 the golden rule of the, 32.
 the herb, 146, 147.
 the rock, soil for, 113.
 the rock versus the rockery, 111.
 the syllabus of the, 55.
Gardener, his apothegms, 107, 155.
Gardeners, 26, 160, 180, 188.
Gardening, natural style of, 34.
 the art of, 31, 33.
Gardens and privacy, 35.
 to lounge in, 306.
Garden-work in spring, 26.
Gautier, 9.
Geese, wild, 27.
Genista saggitalis, 131.
Gentian, 110, 122.
Gentiana acaulis, 122.
 Andrewsii, 122.
 crinita, 122.

Gentiana verna, 122.
Geranium maculatum, 72.
Gérarde, 155, 289, 314.
Germans as botanists, 214.
Geum, 248.
Giant hyssop, 145.
Gladiolus, 302.
Göthe, 208.
Golden-banded lily. See L. AURATUM.
Golden bell. See FORSYTHIA.
Golden-rod, 305, 312.
Goldfinch, 312.
Goldthread, 77, 78, 127, 255.
Grackle. See CROW-BLACK-BIRD.
Grape hyacinth, 146.
Grass, crab, 44.
Grasses, lawn, 38.
 ornamental, 37.
Grasshoppers, 204, 251, 279, 307, 320.
Gray, Asa, 54.
Gray, David, 331.
Gray's elegy, 69.
Great groundsel, 248.
Herball, 315.
 sphinx, the, 220–223.
Green leaf cricket. See CRICKET.
Ground pine, 78.
Guelder rose, 230.

Habenaria blephariglottis, 129.
 ciliaris, 129.
 fimbriata, 129.
 psychodes, 129.

Hair-bird, 193.
Halesia tetraptera, 232.
Hamamelis Virginiana, 337.
Hamerton, P. G., 75, 95.
Harebells, 131.
Harpalium rigidum, 294.
Harris, Thaddeus W., 21.
Hart's-tongue fern, 78.
Hawk-moths, 220, 221.
Heleniums, 294.
Heliantheæ, 256, 280, 287-295, 301, 305, 317.
Heliopsis lævis, 257, 288, 292.
Hellebore, black, 16.
Hemerocallis flava, 138, 142-144, 150, 170.
   fulva, 142, 144.
   graminea, 144.
   kwanzo variegata, 144.
   kwanzo variegata, fl. pl., 144.
Hepaticas, 59, 62, 72, 106, 107, 110, 116, 254, 256.
Herbalists, the old, 214, 290.
Herrick, 276, 335.
Herrick's Julia, 87.
Hibiscus, 234.
Hickory, 327.
Hieraceum aurantiacum, 160.
Hildreth, Charles Lotin, 13.
Hollyhock, 146, 159.
Holmes, O. W., 70, 161.
Honey-bees, 199, 204, 212, 213, 333.
Honeysuckles, climbing, 239.
   Japanese golden-leaved, 241.
   Japanese, or Halleana, 197, 206, 220, 222, 240, 251, 275.
Honeysuckles, monthly fragrant, 241.
   shrubby, 36, 146, 235.
   swamp, 255.
   wild, 137.
Hood, Thomas, 205, 332-334.
Hoop-petticoat daffodil, 90, 98.
Horace, 63.
Hornbeam, 255, 337.
Hornets, 202-204.
Horsemint, 249.
Hoteia Japonica, 154.
Howells, 273.
Hudsonia tomentosa, 119.
Humming-birds, 174, 206, 220, 223.
Hunt, Leigh, 333.
Hydrangea, 35.
   paniculata grandiflora, 235, 280, 304, 306.
Hylodes, 19.
Hypoxis erecta, 125.

Iberis corifolia, 124.
   correæfolia, 124.
   Gibraltarica, 124.
   jucunda, 125.
   sempervirens, 125.
   tenoriana, 125.
Indian summer, 331.
Insect fertilization. Chapter IX.
   music, 204.
   pests, 160, 171, 186, 188, 203. 204, 217, 239, 282, 295, 304, 319.

Insecticides, 45.
Insects, edible, 22.
  injurious to vegetation, 21, 319.
Ipomœa, 15.
Iris cristata, 124.
  English, 138.
  germanica, 137, 139, 146, 148.
  Kœmpferi, 137.
  pumila, 124.
  reticulata, 124.
  Spanish, 138.
  Susiana, 138.
  versicolor, 72.

Jack-in-the-pulpit, 72.
Jefferies, Richard, 116, 134, 244, 297.
Jerusalem artichoke, 317.
Jesse, Edward, 13, 201, 202, 329.
Jewel-weed, 252, 256.
Jonquils, 6, 89, 97, 117.

Kalmia latifolia, 233.
Katydid, 19, 280.
Keats, John, 120, 161, 332-334.

Ladies-tresses, 129, 256.
Lady's-slippers, 128, 183.
Lamb, Charles, 309.
Lang, Andrew, 60, 192, 194.
Larkspur, 146, 161, 206, 245, 250, 300.
Lawn, the, 38-46.

Lawn, pests of the, 43.
Leiophyllum buxifolium, 119.
Lemon-balm, 145.
Lemon-verbena, 250.
Lichens, 50.
Life in the country, 147.
Ligustrum, 232.
Lilac, 36, 136, 146, 229.
Lilies, 37, 49, 89, 161, 165, 195, 246.
  among ferns, 185.
  synopsis of, 169.
  transplanting, 173.
Lilium auratum, 5, 7, 168, 177, 275, 303.
  Brownii, 176.
  bulbiferum, 26, 177.
  Canadense, 141, 169, 183.
  Canadense flavum, 183, 185.
  Canadense rubrum, 141, 170, 182.
  Canadense, varieties of, 184.
  candidum, 172, 174.
  Chalcedonicum, 174, 206.
  colchicum, 170, 171.
  croceum, 170.
  excelsum, 174.
  giganteum, 180.
  Grayi, 181.
  Hansoni, 174.
  Harrisii, 176.
  Humboldtii, 175.
  Isabellinum. See EXCELSUM.
  longiflorum, 176.
  martagon, 177.

Lilium, martagon album, 177.
  martagon dalmaticum, 177.
  pardelinum, 175, 176.
  pardelinum Alpinum, 176.
  Parryi, 175.
  parvum, 176.
  Philadelphicum, 170, 172.
  pulchellum, 172.
  rubescens, 175.
  speciosum, 168, 176.
  speciosum Melpomene, 176.
  superbum, 141, 181, 182.
  tenuifolium, 171.
  testaceum. See EXCELSUM.
  Thunbergianum, 177.
  tigrinum, 181, 275.
  tigrinum fl. pl., 181.
  tigrinum splendens, 181.
  umbellatum, 177.
  Washingtonianum, 175.
Lily, legend of the, 166.
  Madonna, 166, 172, 174.
  of the valley, 6, 146.
  tiger, 146, 181, 275.
  wild Turk's-cap, 181.
  wild wood, 183, 198, 206, 256.
Lime-tree, the, 193, 198, 199, 206, 230, 333.
Linaria vulgaris, 312.
Linden, 319. See also LIME-TREE.
Linnæa borealis, 77, 111, 125, 127.
Lobelia, blue, or syphilitica, 252.

Lobelia, cardinalis, 252.
Locust, 204.
Lonicera Halleana, 193, 206, 220, 222, 240, 275.
Lotus corniculatus, 131.
Lowell, J. Russell, 110, 336.
Lubbock, Sir John, 209.
Lungwort, 72.
Lychnis, scarlet or Chalcedonica, 206, 245, 248.

Madonna lily, 166, 172, 174.
Magenta, 5, 15, 51, 55, 61, 152, 153.
Magnolias, 26, 36, 83.
  conspicua, 84.
  glauca, 85.
  Halleana, 36, 84, 240.
  Magnolia Lennei, 84.
  macrophylla, 84.
  purpurea, 84.
  Soulangeana, 85.
  Thomsoniana, 85.
  tripetala, 85.
Maiden-hair, 327.
  fern, 261, 268.
Maple, Japanese, 235, 327.
  Norway, 327.
  scarlet, 27, 28, 326, 327, 331.
  sugar, 327, 331.
Marigold, 276.
Marsh marigold, 75, 76, 137.
Martagon lily, 177, 198.
Martial, 250.
Martin, purple, 18.
Martin, Sir Theodore, 101.
Matthiola annua, 303.

May-beetle, 45, 282.
May-flies, 149.
Meadow-lark, 27, 59, 82, 321.
Meconopsis Cambrica, 120.
   Nepalensis, 120.
Meleager, 251.
Menispermum Canadense, 242.
Mertensia Virginica, 72.
Midges, dance of the, 205.
Midsummer - Night's Dream, the, 193.
Mignonette, 250.
Migrating birds, 320.
Milkweed, 160, 252, 320.
Miller, Hugh, 24.
Missouri currant, 87.
Mitchella repens, 126, 127.
Mock-orange, 36, 146, 148, 231.
Monarda, 206, 249.
Monk's-hood, 31, 146, 338.
Montgomery, Alexander, 164, 175.
Moon-flower, 15.
Moonseed, 242.
Moonwort, 267.
Mosquitoes, 204.
Moss-pink. See PHLOX SUBULATA.
Moths, 219, 223, 224.
Mottoes, 187, 197.
Mountain-ash, 319.
Mourning-dove, 193.
Mulleins, 257, 312.
Müller, Hermann, 209, 214, 216, 218, 221.
Mushrooms, 307, 308, 312.

Music, heating *vs.* cooling, 197.
Mutton, a roast leg of, 311.
Nankeen lily, 174.
Narcissus. See also DAFFODIL.
   Ard Righ, 94.
   cernuus, 95.
   corbularia, 98.
   corbularia citrina, 90.
   double poeticus, 94, 97, 144.
   emperor, 94.
   empress, 92, 93.
   Horsfieldi, 4, 87, 92, 93.
   incomparabilis, 91, 93.
   incomparabilis Barri, 91.
   incomparabilis Leedsi, 91.
   incomparabilis Nelsoni, 91.
   Leedsi Circe, 95.
   Leedsi cynosure, 96.
   Mary Anderson, 96.
   maximus, 94.
   nobilis, 95.
   obvallaris, 4, 95.
   odorus, 6.
   orange phœnix, 93, 98.
   pallidus præcox, 4, 95.
   paper-white, 6.
   poeticus, 93, 97, 117.
   princeps, 95.
   pseudo-narcissus, 93, 96.
   Sir Watkin, 94.
   sulphur phœnix, 93, 98.
   tazetta, 6, 97, 99.
   tazetta, grand primo, 6.
   tazetta, grand soleil d'or, 6.
   Telamonius, 95.

Nasturtium, 276.
Necessity of the hour, a, 17.
Nicotiana affinis, 286.
  tabaccum, 286.
Night-flowering stock, 224.
Nightshade, purple, 252.
Noli me tangere, 256.
Norway maple, 327.
Nosegay, the old-fashioned, 249.
Nurserymen as money-makers, 83.
Nut-hatch, 255.

Oak, scarlet, 328.
Oak-fern. See FERNS.
Oaks, 318, 327.
Ode, an autumnal, 320.
Odors, 3, 6–9, 51, 136, 140, 145, 146, 151, 175, 186, 220, 231, 232, 242, 252, 275, 285, 286, 303, 338.
  attractive to insects, 218–220, 224.
Œcanthus fasciatus, 279.
Œnothera biennis, 286, 287.
Onoclea sensibilis, 266.
  struthiopteris, 183, 264.
Onosma taurica, 131.
Orange-lily, 165, 170.
Orchids, 7, 88, 127.
Orchis foliosa, 130.
  latifolia, 130.
  maculata, 130.
  spectabilis, 77, 128.
Oriental poppy, 150, 170.
Oriole, 18, 82.

Orobus vernus, 160.
Osmunda cinnamomea, 266.
  Claytoniana, 266.
  gracilis, 265.
  regalis, 265.
Ostrich fern. See FERNS.
Ostrowskia magnifica, 157.
Owls, 12, 13, 222, 253, 254.
Ox, the phantom, 195.
Oxlips, 4.

Pæonias, 37, 138, 139.
Papaver Alpinum, 120.
  Hookeri, 122.
  nudicaule, 120, 121.
  Orientale, 150, 170.
  Orientale bracteatum, 150.
  Parkmanii, 150.
  umbrosum, 121.
Partridge-vine, 78, 111, 126, 127, 198, 255.
Pasque-flower, 123.
Peach, Japanese double-flowering, 86.
  red-flowering, 86.
  rose-flowering, 86.
  versicolor plena, 86.
Peepers. See HYLODES.
Pepperidge-tree, 328.
Perfume. See ODORS.
Periploca græca, 242.
Petunia, 220, 276.
Pewee, wood, 159, 277.
Phantom ox, the, 195.
Phegopteris dryopteris, 269.
Philadelphicus coronarius, 231.

Index. 351

Philadelphicus foliis aureis, 231.
Gordonianus, 231.
Phlox amœna, 119.
 divaricata, 119.
 perennial, 37, 118, 146, 280, 281, 303.
 procumbens, 119.
 subulata, 117, 119.
Phœbe-bird, 277.
Pink lady's-slipper, 128, 183.
Plant infanticide, 118.
Planting too closely, 85.
 too sparingly, 320.
Plants, Californian, 158.
 capricious, 132.
 carpet, 125–127, 130.
 deterioration of, under culture, 127.
 fertilized by insects, 141.
 half-hardy, 53.
 hybridizing, 140, 142, 281.
 ill adapted to climate, 233.
 massing, 52.
 staking, 157, 203.
 to be avoided, 52, 115, 131, 144, 268, 284, 285, 294.
 to propagate, 49, 50.
 transplanting, 48, 49, 173, 320.
 water, 76, 137.
Plover, 27.
Plum, double-flowering, 84.
 purple-leaved, 36, 235.
Plume-poppy, 284.
Polygala lutea, 119.
 paucifolia, 77.

Polygonum cuspidatum, 284, 285.
Polypodium falcatum, 269.
Polypody. See FERNS.
Pope, Alexander, 305, 322.
Poppy, Alpine, 120.
 double white, 146.
 Iceland, 52, 120, 121, 171, 250.
Potash and iron medicine, 24.
Pot-pourri, 10, 11.
Primroses, 4, 107, 116, 309.
 in poetry, 109, 110.
Primula auricula, 108, 116.
 cortusoides, 108.
 denticulata, 108.
 farinosa, 109.
 Mistassinica, 109.
 Parryi, 109.
 polyanthus, 108, 116.
 rosea, 108.
 Sieboldi, 108.
 Sikkimensis, 108.
Prince's pine, 78.
Privets, 232, 236.
Protection, advantage of, for trees, 84.
 for plants, 114, 115.
Pruning, cruelty of, 83.
Prunus, 36.
 triloba, 84.
Pteris aquilina, 267.
Pyrethrum, 152, 316.
Pyrolas, 78, 127.
Pyrus Americana, 319.
 aucuparia, 319.
 malus, 86.

Quince, Japan, 35, 85.
Ranunculus aquatilis, 76.
Read, Thomas Buchanan, 336, 337.
Red spider, 27, 304.
Redwing, 27.
Rhododendron, 139, 140.
Rhus glabra laciniata, 329.
Ribbon-grass, 146.
Ribes, 36.
Robin, 4, 16, 18, 159, 319.
Robinson, Phil, 16, 17.
Rocket, double-flowering, 146, 161.
Romneya Coulteri, 158.
Rook, the, 15.
Rose-beetle, 171.
Rose, brier, 146.
   Christmas, 16.
   legend of the, 186.
   Marie Rady, 189.
   of Sharon, 35, 146, 234.
   pests, 188, 203.
   pot-pourri, 10, 11.
   wild, 252, 305.
Roses, 7, 158, 161, 165, 186, 239, 246, 276.
   autumnal flowering, 276.
   species, hybrids, and varieties of, 187-190.
Royal fern. See FERNS.
Rudbeckia, 288-293.
Rue anemone, 73, 110, 117.
Ruffed grouse, 253, 267.

St. Peter's-wort, variegated, 236.

Salvias, 325.
Sambucus, 231.
Sand-myrtle, 119.
Sanguinaria Canadensis. See BLOODROOT.
Saponaria ocymoides, 131.
Sassafras, 147, 327.
Savarin, Brillat, 309.
Saxifraga cordifolia, 116.
   cotyledon, 117.
   longifolia, 117.
   peltata, 117.
Scabiosa atropurpurea, 248.
   Caucasica, 248.
Scarlet maple, 27, 28, 326, 327, 331.
Scillas, 106, 107, 116.
Scolopendrium vulgare, 270.
Sea-lavender, 249.
Sea-shore, the, 197.
Sedum, 111, 112, 115.
Senecio macrophylla, 248.
   pulcher, 248.
Sensitive fern. See FERNS.
Sentiment, a charming, 17.
Shad-blow, 73, 328.
Shakespeare, 2, 13, 110, 200, 244, 323.
Shrubs, autumnal coloring of, 326, 329, 331.
   hardy border, 35, 37, 229-236.
   pruning, 36, 37.
   the sweet-scented, 35.
   with dark foliage, 235.
   with ornamental fruit, 236, 301, 317, 318.

Shrubs, with variegated foliage, 235.
Silene, 131.
  alpestris, 125.
Silk-vine, 241.
Silphiums, 293, 294, 305.
Silver-bell, 232.
Skunk-cabbage, 60.
Smilacina bifolia, 125.
Snails, 116.
Snake-head, 252.
  root, 318.
Snowball, 36, 146, 230.
Snowberry, 236.
Snow-drops, 3, 31, 110, 146.
Snow-pink, 146, 155, 156.
Soils, colors of, 61.
  treatment of different, 46, 47, 304.
Soldanella, 111.
Sounds, 8, 19-21, 27, 28, 105, 106, 159, 199, 204-206, 209, 250, 251, 253-255, 277-280, 319-322, 332, 336, 337.
Sour-gum, 328.
Sparrow, English, 16-18, 20, 43, 81, 149, 159, 188, 296.
  song, 18, 27.
  tree, 255.
  white-crowned, 81.
  white-throated, 81, 254, 320.
Speedwell, 249.
Sphinges, the, 220-223.
Sphinx, Carolina, 220.
  cinerea, 220.
  drupiferarum, 220.
Spiders, 295.

Spiranthes cernua, 129, 300.
  gracilis, 129.
Spiræa aruncus, 154.
  filipendula, 154, 171.
  Humboldtii, 154.
  lobata, 154.
  palmata, 155.
  prunifolia, 86, 329.
  Thunbergii, 86.
  ulmaria fl. pl., 154.
Spiræas, herbaceous, 154.
  shrubby, 36, 229, 230, 329.
Sprengel, Christian, 214.
Spring, an early, 27.
  beauty, 59, 62, 110, 116, 256.
  bitter-vetch, 160.
Squirrels, 12, 254, 320, 333.
Star-flower, 77, 78, 127.
Star-grass, 78, 125, 127.
Starling, 321.
Starworts, 305, 314, 315.
Statice, 249.
Stinging annoyances, 203, 256.
Stocks, 146.
  double ten-weeks, 303.
  night-flowering, 224.
Street, Alfred Billings, 336.
Sugar-maple, 327, 331.
Sumac, 329, 331.
  cut-leaved, 36, 329.
Sun-dial, the, 297, 306, 307.
Sunflower, Gérarde's description of the, 290.
Sunflowers, 146, 287-295.
Swallows, 147, 193, 205, 220, 333, 336.

Swallows, chimney, 19.
Swamp, a, 252.
  honeysuckle. See AZALEA VISCOSA.
  pink. See AZALEA NUDIFLORA.
Sweet birch, 255.
  pepper-bush, 232.
  sultan, 248.
Sweet-gum, 326, 328.
Sweet-william, 146, 155, 156, 250.
Sword-grass, 146.
Symonds, John Addington, 117, 187.
Syringa, 35, 231.
  golden, 236.

Talmage, Dr., 5, 203.
Tanacetum crispum, 285.
Tansy, curled-leaved, 285.
Tecoma radicans, 241.
  radicans var. atrosanguinea, 241.
Tennyson, 13, 159, 161, 199, 221, 332.
Thalictrum anemonoides, 73.
Theocritus, 202, 251.
Thomson, James, 193, 335.
Thoreau, 13, 58, 148, 251, 260.
Thorns, 74, 313, 319.
Thrush, 8, 159.
  hermit, 81.
  wood, 81, 82.
Thyme, 146, 285.
Tiarella, 117, 126.
Tiger-lily, 146, 170, 181, 275.

Tilton, Theodore, 195.
Toad-flax, 312.
Toads, 105, 106, 115, 116.
Tobacco-plants, 286.
Touch-me-not, 256.
Tradescantia, 146.
Transplanting, best season for, 330.
Tree-sparrow, 255.
Tree-toad, 193.
Trees, autumnal hues of, 325-330.
Trillium, 62, 64, 65, 75, 254, 256.
  erectum, 64.
  erythrocarpum, 64, 117.
  grandiflorum, 117.
Tritoma, 280, 302.
Trout-fishing at night, 221.
Trout, Rocky Mountain, 18.
  speckled, 18, 254, 307.
Trumpet-flower, 147, 241.
Tulips, 4, 146, 151.
Tunica saxifraga, 125, 300.
Turk's-cap lily, 181, 198.
Turner, Charles Tennyson, 335.
Twin-flower, 77, 111. See also LINNÆA BOREALIS.

Uvularia grandiflora, 73.

Vaccinium, 77, 127.
Valerian, 146, 160.
Veery, 253, 321.
Verbena, 220, 276.
Verdant sculpture, 37.

Veronica longifolia subsessilis, 249.
pumila, 125.
repens, 125.
rupestris, 125.
verbenacea, 125.
Vervain, 256.
Vesper-sparrow, 161.
Viburnum lantanoides, 236, 318.
opulus, 236, 318.
opulus sterilis, 230.
plicatum, 231.
rugosum, 318.
Viola blanda, 66.
cornuta, 300.
cucullata, 65.
odorata, 66.
pedata, 65.
pedata bicolor, 65, 300.
rostrata, 119.
Violets, 5, 7, 8, 62, 63, 65, 66, 109, 110, 161, 256.
a hot-bed of, 11, 12.
how they became purple, 66.
Marie Louise, 6.
Shakespeare's, 66.
under the, 70.
Virgil, 250, 315.
Virgin's bower, 237, 238, 239, 252.
Voices of Nature, 204. See also SOUNDS.
winter, 16.

Walton, Izaak, 201.
Wasps, 202, 204.

Water-lily, 5, 75, 106.
Water-plants, 76, 137.
Water ranunculus, 75.
Wayfaring tree, 318.
Weather, the, 14, 15, 22, 23, 27, 28, 135, 193, 203, 245, 246, 306.
Weeds, the big, 256.
Weigela, 37, 234, 235.
white, 36.
White alder, 74, 232.
fringe, 230.
White, Gilbert, 116, 205, 329.
White-crowned sparrow, 81.
White-throated sparrow, 81, 254, 320.
Whittier, J. G., 110, 194.
Wild bean, 242.
carrot, 312.
rose, 252, 305.
thorns, 74, 313.
Turk's-cap lily, 181.
Wild wood-lily, 183, 198, 206, 256.
Wind, Ruskin's plague, 23.
Wind-flower. See ANEMONE.
Winter-green, 78, 126, 127, 255.
Wistaria, 236.
Witch-hazel, 337.
Woman as a monopolist, 8, 9.
Woodcock, 27, 252.
Woodpecker, downy, 255.
Wood-pewee, 159, 277.
Wood-thrush, 8, 159.
Woods, the wintry, 255, 267, 337.

Woodsia Ilvensis, 270.
  obtusa, 270.
Wordsworth, 96, 187.
Wren, 18.

Xylocarpa Virginica, 210, 211.

Yellow-bird, 320.
Yellow-wood, 230.
Yew, 331.

Zenachus, 250.
Zinnia, 5, 15, 316.

www.ingramcontent.com/pod-product-compliance
Lightning Source LLC
Chambersburg PA
CBHW031425230426
43668CB00007B/433